A Photographic Guide to North American Raptors

A Photographic Guide to North American Raptors

Brian K. Wheeler
William S. Clark

ACADEMIC PRESS

San Diego London Boston New York Sydney Tokyo Toronto

AP Natural World is published by
ACADEMIC PRESS
525 B Street, Suite 1900, San Diego
California 92101-4495, USA
http://www.apnet.com

ACADEMIC PRESS
24–28 Oval Road
London NW1 7DX
http.//www.hbuk.co.uk/ap/

This book is printed on acid-free paper.

A catalogue record for this book is available from the British Library

ISBN 0-12-745531-0

Typeset by Goodfellow and Egan Ltd, Cambridge

Printed in Great Britain by Butler and Tanner, Frome, Somerset

99 00 01 02 03 BT 9 8 7 6 5 4 3 2 1

CONTENTS

FOREWORD

You hold in your hands the key that unlocks the heavens—or more accurately—the birds that soar there. It may look like just another field guide, but it is more, and I stand first in line to tell you this. As a hawk watcher in long standing, I have eagerly awaited this utilitarian partner to Clark and Wheeler's *Hawks*. First, because as a reference tool, it will be invaluable to serious hawk watchers. The information it contains is age class and subspecies specific; the format is straightforward; and the material is easily referenced.

But experience has not made me forget my own early efforts to gain proficiency in the art of raptor identification. This new guide, with its employment of photos and integrated concern for the problems presented by lookalike species, will also serve the novice, saving them much time, effort, embarrassment, and...

Frustration! You who are fortunate enough to have this guide in your hands cannot *begin* to imagine how frustrating it was trying to learn the rudiments of raptor identification twenty years ago. The guides available then were the standard field guides—well suited for identifying perched songbirds; ill suited for distinguishing wind borne birds of prey or distant perched raptors.

The text in these generalized guides was necessarily terse and occasionally oblique. Observers might be directed to search for field marks that were invisible at any appreciable distance and even to search for field marks that did little to distinguish one species from another. Small wonder there was frustration. The illustrations, too, in the standard guides fell short in the hawk watching arena. Most were overdrawn and not a few were anatomically incorrect.

In my neophyte days, I used to sit on a strategic ridgetop with those popular field guides spread in a semi-circle around me. I would study each passing bird of prey and then strive to find the bird's likeness replicated in the books. Failure was my companion.

I recall one illustrative occasion at Hawk Mountain Sanctuary in Pennsylvania. I was conducting the official count, filling in for the official counter whose estimates of my skills were flattering but misplaced. Among the many sharp-shinneds, several broadwings and sundry other raptor species seen that day was a pair of unrecognizable soaring raptors that shed any name experience or erudition could bring to bear. The pair went down in the books as "Unidentified."

You, however, will have no difficulty finding the photo likeness of my mystery birds replicated in this guide. All you have to do is flip to the chapter on Peregrines and study the profile of a soaring bird. Why did the identification confound me at the time? Because soaring Peregrines were nowhere depicted in those generalized field guides of the day. If not perched, falcons were invariably depicted in level flight, in fact were *distinguished*, according to many authorities, by their tendency *not* to soar. Because of this counsel, I did not consider the possibility that my birds may have been falcons (and I did not identify the birds).

Anyone who has spent **any** time at migratory junctions like Cape May, New Jersey can tell you that falcons soar well and frequently. On a soaring Peregrine the acutely angled wings straighten, the harsh lines soften. The tail fans, masking the length of the wing, making the birds seem deceptively buteo-like.

Set against the backdrop of a cloud-pocked sky, a soaring Peregrine exhibits a dreamlike quality so at odds with the flash and dash of the birds in point to point flight. It was, in fact, in Cape May, beneath a falcon studded sky that I met William S. Clark, the coauthor of this book. That was the autumn of 1976.

Bill was the founding director of the New Jersey Audubon Society's ambitious new Cape May Bird Observatory. I was the first "official counter," my task, to identify and record each and every passing bird of prey. Our meeting, in the grassy parking lot behind the "North [hawk banding] Station" was fortuitous. Friendship and shared interest has been our bond. That raptor-filled fall was the most memorable of my life and its imprint is enduring. Though I have traveled widely, even moved to northern New Jersey for a time, I have always returned to Cape May, the place that has been dubbed "The raptor capital of North America."

Bill's orbit, too, has expanded beyond Cape May. His peregrinations in search of the world's birds of prey are legendary and his travels have taken him to places that even Rand McNally has never heard of. On a global scale, there are few living human beings who know more about birds of prey than William S. Clark and none, to my mind, who know more about their identification. On the more focused subject of North America's diurnal raptors, he has few peers—but one of them is Brian K. Wheeler, whose photos dominate this book.

If you sense artistry in this book's photography, that is because Brian Wheeler is an accomplished artist, as adept with a brush as he is with a telephoto lens. He is also a familiar figure at those famous hawk-watching junctions scattered across North America—the Rio Grande Valley where migrating hawks rise in swirling tornadoes... Duluth, Minnesota where Goshawks fly beak to tail... Holiday Beach, Ontario where Broad-winged Hawk numbers darken the sky... Hawk Mountain, Pennsylvania where sixty years ago, it all began.

There are many nature photographers who specialize in capturing birds of prey on film. Few are as good as Brian and some say none. Getting WOW photos of raptors is one thing, but assembling a collection of 375 plus photos that thoughtfully and accurately depict North America's birds of prey is quite another.

An effort such as this guide takes guided purpose. It requires that photos be filtered through a fine, calculated mesh of experience and purpose. For every photo you see here, there were a hundred discards and a dozen seconds that did not make the cut. Photos that did not show the angle of a wing quite right...did not illustrate the chest pattern of an immature bird properly...did not show what Wheeler and Clark wanted to show.

And you can trust the likes of Bill Clark and Brian Wheeler to know just what plumage and anatomical points need to be shown. Though they are experts, they have not forgotten the trials and pitfalls that mark the paths of the beginner. This book proves it.

So now that you have the key that unlocks the heavens, go out and watch some hawks. Wheeler and Clark have gone a long way toward taking the frustration out of the endeavor, leaving more time for the excitement and wonder that defines hawk watching.

Hats off to them. Good hawk watching to you.

Pete Dunne
Director, Natural History Information
Director, Cape May Bird Observatory
New Jersey Audubon Society

ACKNOWLEDGMENTS

We received a great deal of help from a host of friends and colleagues during the preparation of this guide; they are mentioned below. We are especially appreciative of the friendly assistance, encouragement, and indulgence given to us by our Academic Press editor, Andrew Richford.

The following, in alphabetic order, assisted us in our field work or provided a base from which to operate or both. They are: Bill Alreth, Bud Anderson, Sharon Bartles, Eddie Bigelow, Pete Bloom, Sue Calley, Jack & Dolly Clark, Jane Church, Jeff & Joan Dodge, Tom & Andrea Doolittle, Dudley & Nancy Edmondson, Dave & Molly Evans, Rod Gray, Ned & Linda Harris, Joe & Elaine Harrison, Steve Hoffman, Paul James, Tammy James, Alan Kleir, Tony Miller, Frank Nicoletti, Lynn Oliphant, Bob Pantle, Chuck Preston, Nancy Reid, Steve & Ruth Russell, Paul Saunier, Mike & Noreen Sankovich, John Schmitt, Jay & Ginny Schnell, Chris & Brenda Schultz, Mitch Smith, Kim Stahler, Dave Tetlow, Ian Warkenton, Loren & Kim Wheeler, Bob Yunick, Karl Zacharias, and Louise Zemaitis.

Many photographers kindly permitted us to inspect their raptor slides as possible candidates for the guide. Those whose slides were not used are: Simon Bijlsma, Alan Jenkins, Larry Sansone, Paul Schnell, and Monte Taylor. Their slides were not used, not because they were lacking in any way, but simply because one of us had taken an acceptable photo of the same subject. Those whose slides were included in the guide are Dave Clendenen, Dudley Edmondson, Jim Gallagher, Ned Harris, Russ Kerr, Ken Knowles, Bruce MacTavish, Dick McNeely, Jim Parnell, Ray Schwartz, Noel and Helen Snyder, and Russell Thorston. In addition, the Frank Lane Picture Agency provided one photo by Mark Newman, and VIREO provided a photo each by Frank Haas, Franz Lanting, Art Morris, Van Remson, and Dale Zimmerman. The photo or photos provided by each of them are listed in the Photo Credits.

We are indebted to the Cape May Bird Observatory, the Denver Museum of Natural History, the Bird Division of the Smithsonian Institution, and Trans-Western Express for support during our research.

We saved our biggest THANK YOU for last. This guide would not have been possible if it were not for the support and encouragement of Ellie Clark and Loren and Kim Wheeler.

Brian Wheeler
Bill Clark

PHOTO CREDITS

INTRODUCTION

Our goal in preparing this photographic guide for North American diurnal raptors was to show, using color photographs, every recognizably different plumage of each species of regularly occurring diurnal raptor, as well as representative plumages of vagrant and extraterritorial raptor species. Herein we present 377 photographs together with the field marks that will enable an observer in the field to identify our diurnal raptors as to species, and further to be able, in many cases, to determine their age, sex, race, and color morph. It was also our goal that, by including only the highest quality of photographs, this guide would also be an aesthetic work of art.

We did not intend that this guide would be a complete field guide but that it would complement our raptor field guide, *Hawks*, which was published by Houghton Mifflin in the Peterson field guide series.

We originally planned that the two of us would take all of the 360 plus photographs that were proposed for the guide. However, we found that this was just too large a task, and we have therefore included 24 photographs taken by other photographers to get the complete coverage we had envisioned. The other photographers are listed in the photo credits and Acknowledgments.

This was a team effort, with both of us having different but overlapping and shared responsibilities. Bill arranged for the contract with Academic Press and was responsible for the preparation of the text and captions. Brian made the final slide selection and designed the layout of the accounts. Nevertheless, we both had a considerable amount of input into the other's areas. Both of us spent much time in the field taking photographs; Brian's quota was three-fourths of them, and Bill's was one fourth. Brian's specialty area was the buteos and eagles, and Bill's was the overseas raptors. We both worked with the publisher's staff on the layout of the guide, especially on the species accounts and Identification Problems section.

We began our collaboration on raptor identification many years ago and produced *Hawks*, as mentioned above. Since its publication, we have both continued working in the field observing and studying raptors to test the known field marks and accepted hawk ID lore. As a result, we have found many new field marks and pinpointed others that do not work. All of this new information is presented herein.

Every species of raptor that has been recorded in the United States and Canada is covered in this guide, even if there is only a single record. We included the extra-limital Hawaiian Hawk, as Hawaii is one of the United States (but not in North America). The Aplomado Falcon, extirpated from its former breeding range in parts of New Mexico, Arizona, and Texas, is now being reintroduced in Texas, with plans to do so elsewhere in the Southwest.

We do not show all plumages on vagrant species, only the ones most likely to be seen. There are single records of the Crane Hawk and Collared Forest-Falcon from southern Texas on the Mexican border. The other vagrants, Roadside Hawk, Steller's Sea Eagle, White-tailed Eagle, Northern Hobby, and Common Kestrel, all have been recorded more than once but fewer than a dozen times.

Photographic equipment and techniques

Since we began our collaboration on raptor field ID, Brian has been upgrading his photographic equipment to get the best possible photographs. We felt and still feel that a lens with a 400 mm focal length is best overall for taking photographs of flying and perched raptors. With that in mind, Brian purchased a Nikkor ED 400 mm f/3.5 lens in 1985 and immediately began taking higher quality photographs. Bill followed his lead and purchased one of these lenses in 1988. All of our photos used in this guide were taken with this wonderfully sharp, manually focused lens. Some perched photos were taken with a Nikkor 1.4B teleconverter attached. Brian has used Nikon F3 bodies for many years but recently switched to Nikon F4's. Bill had been using a Nikon FE2 until several years ago, when he switched to a Nikon 8008. All cameras used had motor drives attached or integral.

Both of us have used Kodachrome 64 for many years, as this was the sharpest film available until recently. Bill began experimenting a few years ago with various FujiChrome films and found that Velvia was just as sharp as Kodachrome 64 with better color saturation. Late in the project, we both used Velvia (ISO 50) pushed one stop to ISO 100. Bill, however, had also been rather satisfied with the new Kodak Lumiere 100. We both have recently switched to FujiChrome Provia and are extremely satisfied with it. Most of the photos in the guide were taken with Kodachrome 64, but many are Velvia and a few are Lumiere 100X and Provia.

We purposefully stayed away from photographing raptors at nests or in staged or studio settings. Some of our photos were taken from blinds, but most were taken in the open or using a car as a blind. Some shots were taken using a tripod or the car window for support. We have specialized in taking flight shots while hand holding camera and lens. Bill uses a Bush Hawk shoulder stock, but Brian uses just the camera and lens.

The requirements for any color photograph to be considered for the guide was that it was taken with slide film and is sharp, well lighted, fairly close, and depicts an age, sex, color morph, or race different from already accepted photographs.

Organization of guide

The main section of the guide consists of an account for each species, presented in standard taxonomic order, but with the vagrants following the regularly occurring species. Each species account starts with a short text that briefly describes the general field marks of that species, the characters of each age, sex, color morph, and race, and a brief description of each recognizably different plumage. Next is a list of similar species and the field marks to distinguish them. This is followed by a short description of the range of the species in North America.

In the next section are presented measurements of length, wingspan, and weight, all of which were taken on live birds or from museum specimens. No data were repeated from published sources, as some are erroneous. These data are given in both metric and English units. Measurement data are not easy to use, as judging size in the field is extremely difficult. But they can be used to compare the relative sizes of raptors that are flying or perched together.

The main part of each species account is the set of color photographs. For each regularly occurring species, the set includes a photo of each different plumage. Each photo has a caption below it that gives the age, sex, color morph, or race of the subject raptor, as well as the field marks shown. We decided not to use Peterson style arrows to point to field marks, as they might detract from the aesthetic quality of the photos.

Following the species accounts is a section entitled "Raptor Identification Problems." This is composed of 14 of the most difficult, in our opinion, North American raptor ID problems. In this section, selected photos, all of which were used in the species accounts, are placed next to photos of similar species so that a direct comparison can be made by the readers. Captions for these photos are different from the captions in the species account.

Age terminology. We have followed the American Ornithologists' Union (AOU) and the American Birding Association (ABA) in adopting the so-called Humphrey–Parkes terminology (see Humphrey & Parkes 1959 *Auk* 76:1–31 and Wilds 1989 *Birding* 21:148–153) for age-related plumage classes, with one exception. The age classes we use herein are Juvenile (juvenal is the adjective), Basic I (etc.), and Adult; these are further defined in the Glossary. The term used for "adult" in the Humphrey–Parkes terminology is "Definitive Basic." We have judged this to be unwieldy for use in a field guide and are using instead the familiar term "adult." The widely used term"immature" has no specific definition other than its literal translation of "not adult" (but see under "immature" in Glossary).

GLOSSARY AND ANATOMY

Adult plumage. The final breeding plumage of a bird. Also called Definitive Basic plumage.

Auriculars. The feathers covering the ears (Fig. 4).

Axillaries. Feathers at the base of the underwing, also called the "armpit" or "wingpit" (Fig. 1).

Back. See Fig. 5.

Band. A stripe of contrasting color, usually in tail.

Barring. See Fig. 5.

Basic I. The second plumage of a bird that takes more than one year to attain adult plumage (e.g., eagles). Usually acquired at one year of age and worn for a year. Basic II and III are following plumages.

Beak. See Fig. 5

Belly. See Fig. 5.

Belly band. See Fig. 1.

Bib. Pattern of uniformly dark breast that contrasts with paler belly.

Breast. See Fig 1.

Buzzard. Name for raptors in the genus *Buteo* and related genera. This name has been mistakenly applied in North America to vultures.

Carpal. The underwing at the wrist, usually composed of all of the primary underwing coverts (Fig. 1).

Cere. A small area of bare skin above the beak (Fig. 5).

Cheek. See Fig. 4.

Collar. A pale band across hind-neck.

Coverts. The small feathers covering the bases of the flight feathers and tail both above and below (Figs 1, 2, and 5).

Crown. Top of head (Fig. 4).

Dihedral. The shape when a bird holds its wings above the horizontal, further defined as:

    (1) Strong dihedral: wings held more than 15 degrees above level.

    (2) Medium dihedral: wings held between 5 and 15 degrees above level;

    (3) Shallow dihedral: wings held between 0 and 5 degrees above level;

    (4) Modified dihedral: wings held between 5 and 15 degrees above level; but held nearly level from wrist to tip.

Dilute plumage. An abnormal plumage in which the dark colors are replaced by lighter, usually creamy color (but not white).

Eye-line. See Fig. 3.

Eye-ring. The bare skin around the eye, somewhat wider in front of the eye in falcons. See Fig. 5.

Face skin. The lores when bare of feathers.

Facial disk. A saucer-shaped disk of feathers around the face, thought to direct sound to the ears.

Feather edge. The sides of a feather. Pale edges usually give the effect of streaking.

Feather fringe. The complete circumference of a feather. Pale fringes usually give a scalloped appearance.

Flank. See Fig. 5.

Fledgling. A raptor that has just left the nest, fledged.

Flight feathers. The primaries and secondaries (Fig. 1).

Forehead. See Fig. 4.

Glide. Flight attitude of a bird when it is coasting downward. The wingtips are pulled back, more so for steeper angles of descent. Its tail is usually closed.

Hackles. Erectable feathers on the nape.

Hawk. A raptor of the genus *Accipiter*, but also used generically for all diurnal raptors.

Hind-neck. See Fig. 3.

Hover. Remain in a fixed place facing into the wind by flapping. More properly called "wind hover."

Immature. Non-specific term which means "not adult." It is used in some field guides to mean "juvenile," in others to mean the stage between juvenile and adult.

Juvenile plumage. First complete plumage, usually different from the adult plumage.

Kite. Remain in a fixed place in moving air on motionless wings.

Leading edge of wing. See Fig. 2.

Leg. See Fig. 5.

Leg feathers. See Fig. 5.

Length. Distance from top of head to tip of tail.

Lores. The area of the face between the eye and the beak (Fig. 3).

Malar stripe. A dark mark on the cheek under the eye (Fig. 3).

Molt. Means by which a bird replaces its feathers.

Monotypic. Having no subspecies.

Morph. Term used for recognizably different forms of a species, usually color related. Color morphs are dark, rufous, and light. See also "phase."

Mustache mark. A dark mark directly under the eye appearing on most falcon species (Fig. 4).

Nape. The back of the head (Fig. 3).

Ocelli. Dark or light spots on the nape and hind-neck which resemble eyes.

Panel. A pale area, usually in the wing, that is noticeable.

Patagial mark. See Fig. 1.

Patagium. The area on the front of the wing between the wrist and the body (adjective is patagial) (Fig. 1).

Phase. Term formerly used for color morph. Phase implies a temporary condition; color morphs are permanent. See "morph."

Plumage. The overall feathering of a raptor, usually replaced annually by molting most or all feathers.

Primaries. The outer flight feathers (Figs 1 and 5).

Primary projection. The distance on the folded wing of perched birds from the tip of the longest primary to the tips of the secondaries.

Raptor. Any bird of prey; any member of the Falconiformes or Strigiformes, although sometimes used to refer only to the diurnal birds of prey.

Rump. The lowest area of the back (Fig. 2).

Scapulars. Row of feathers between back and upperwing coverts of each wing.

Scavenger. Bird that eats carrion, offal, and other decaying material.

Secondaries. The inner flight feathers (Figs 1 and 5).

Shoulder. See Fig. 5.

Soar. Flight attitude of a bird with wings, and usually tail, fully spread. Used to gain altitude in rising air columns.

Spotting. See Fig. 4.

Streaking. See Fig. 4.

Subadult plumage. Plumage of a raptor that precedes adult plumage and appears much like it but with some characters that are not in adult plumage.

Subterminal band. See Fig. 1.

Superciliary line. Contrasting line above the eye (Fig. 3).

Supra-orbital ridge. Bony projection over the eye, giving raptors their fierce appearance (Fig. 5).

Tail. See Fig. 5.

Tail banding. See Fig. 2.

Talon-grappling. Behavior involving two flying raptors that lock feet and tumble with their wings extended.

Trailing edge of wing. See Fig. 2.

Underparts. Breast and belly (Figs 1 and 5).

Undertail coverts. See Fig. 1.

Underwing. The underside of the open wing.

Underwing coverts. See Fig. 1.

Upperparts. Back and upperwing coverts.

Uppertail coverts. See Fig. 2.

Upperwing coverts. Primary and secondary coverts (Figs 2 and 5).

Wing coverts. Feathers that cover the bases of the flight feathers. Three sets are: lesser, median, and greater (Fig. 5).

Wing linings. Underwing coverts. (Fig. 1).

Wing loading. Weight divided by the wing area; a measure of the buoyancy of flight. The lower the wing loading, the more buoyant the flight.

Wing panel. A light area in the primaries, usually more visible from below when wing is backlighted (Fig. 2).

Wingspan. Distance between wingtips with wings fully extended.

Wrist. Bend of wing (Fig. 2).

Wrist comma. Comma-shaped mark, usually dark, at the bend of the underwing; seen on the underwings of many buteos.

FIGURE 1

FIGURE 2

trailing edge of wing

leading edge of wing

wrist

rump

tail banding

undertail coverts

terminal band

underwing coverts

secondary coverts

primary coverts

uppertail coverts

axillaries

subterminal band

breast

patagial mark

belly band

secondaries

flight feathers

primaries

carpal

wing panel

FIGURE 3

superciliary line

lores

eye-line

nape

malar stripe

hind-neck

FIGURE 5

supra-orbital ridge

cere

eye-ring

beak

shoulder

gape

back

lesser coverts

breast

scapulars

barring

belly

flank

median coverts

leg feathers

leg

greater coverts

secondaries

primaries

tail

FIGURE 4

crown

forehead

cheek

auriculars

throat

mustache mark

streaking

Vultures

Vultures are large, bare headed raptors that subsist almost exclusively on carrion, which they tear apart with their strong hooked beaks. Their feet are relatively weak and are not used for grasping prey. Three species occur in North America; two are widespread and the Condor occurs in southern California.

While they are somewhat ungainly on the ground, in flight they are quite graceful.

American vultures often defecate on their legs for cooling or disease control or both.

The misnomer "buzzard" was given to the two smaller American vultures by early European settlers, who thought these birds were related to the darkly colored European *Buteo* with this name. Unfortunately, the name is still in common use.

TURKEY VULTURE *(Cathartes aura)*   Photos TV01–07

The Turkey Vulture, a large brownish-black raptor, has two recognizable plumages: adult and juvenile. Sexes are alike in plumage and size. **Adults** are overall blackish brown with strong iridescence on neck. Head is reddish with an ivory beak. Undersides of flight feathers are silvery and contrast with blackish-brown underwing coverts. Underside of tail is darker silver. Legs are pinkish. **Older juveniles and first plumage adults** (subadults) have pink heads and gradually smaller dark tip to ivory beak. **Juveniles** are like adults except for dusky (not red) head covered with fuzzy down, mostly dark beak (with a small but gradually increasing pale area at base), neat pale edging to feathers on upperparts, and little or no iridescence on neck. In fresh plumage the undersides of their flight feathers are darker than are those of adults.

Best field marks are long, narrow wings, two-toned underwings, long tails, pinkish legs, and red heads (except juveniles). In flight, they are recognized by wings held in a strong dihedral and, in most winds, by their rocking or teetering flight. Wings in strong winds can be held flat or in a modified dihedral.

Similar Species:
Black Vultures (photos BV01–04) are somewhat similar. See under that species for differences.
Golden Eagles (photos GE01–07) can fly with wings in a dihedral and show a two-toned underwing but show a dark band on trailing edge, tawny bands across upper-wings, and golden nape.
Swainson's Hawk dark-morph adults (photo SH07) also fly with wings in a dihedral but have white undertail coverts and lack two-toned underwings.
Rough-legged Hawk dark-morph birds (photos RL06–08, 11–12) also fly with wings in a dihedral and have two-toned underwings, but have dark (adult) or dusky (juvenile) band on trailing edge of wing and on tail tip.
Range: Entire United States, extending barely into southern Canada. Migratory in Northeast and West.
Measurements:
       Length: 62–72 cm (67); 24–28 in. (26)
   Wingspan: 160–181 cm (171); 63–71 in. (67)
     Weight: 1.6–2.4 kg (1.8); 3.5–5.3 lb. (4.0)

TV01. **Adult Turkey Vulture.** Note two-toned underwing; silvery flight feathers contrast with blackish underwing coverts. Red head is visible on birds seen close. Legs reach only halfway along long tail. [TX, Jan]

TV02. **Juvenile Turkey Vulture.** Head and beak of juvenile are dark, otherwise they are like adults, including the two-toned underwing. [FL, Aug]

TV03. **Turkey Vulture seen head-on.** Wings held above horizontal in a strong dihedral. [FL, Jan]

TV04. **Adult Turkey Vulture.** Red head with warts, ivory beak, and iridescence on neck of adults are noticeable. Note pinkish legs and brownish cast to upperwing coverts. Compare head and beak shapes with Black Vulture in photos BV02–04. [FL, Aug]

TV05. **Subadult Turkey Vulture.** Older juveniles, such as this bird, and first plumage adults have pink heads that become progressively redder and ivory beaks with dark tips that become progressively smaller. [TX, Dec]

TV06. **Juvenile Turkey Vulture.** Dark head is covered with fine fuzzy blackish to buffy down. Beak is dark except for small ivory area at base, which gradually expands to about half the beak in the first year. Juveniles show little iridescence on neck. [FL, Aug]

TV07. **Turkey Vulture sunning.** Note pale yellowish areas on uppersides of primaries (compare to whitish areas on Black Vulture in photo BV02). Sunning behavior is common. This bird is an adult. [TX, Oct]

BLACK VULTURE *(Coragyps atratus)*   Photos BV01–04

The Black Vulture, a large black raptor, has two recognizably different plumages: adult and juvenile. Sexes are alike in plumage and size. **Adult** is an iridescent black overall, except for whitish patches in primaries. Head is grayish, with heavily wrinkled skin and yellowish tip to beak. Legs are white. **Juveniles** are similar to adults, but are less iridescent and have black, unwrinkled heads that are covered with fine blackish down. Their beaks are entirely black.

Best field marks are black coloration, short tails, whitish legs, large white primary panels, and powered flight with a series of three to six quick, shallow, almost frantic wing flaps of rigid wings.

Similar Species:
Turkey Vultures (photos TV01–07) are somewhat similar but appear larger and more brownish-black and have red heads (except for juveniles), long, narrow wings with a different underwing pattern, longer tails, and pinkish legs. They fly with their wings in a more noticeable dihedral and their wingbeats are slow on flexible wings, quite different from those of the Black Vulture.
Range: Resident of Southeastern United States from s. New Jersey and Pennsylvania to e. Oklahoma, Texas and very southern Arizona.
Measurements:
        Length: 59–74 cm (65); 23–28 in. (25)
    Wingspan: 141–160 cm (151); 55-63 in. (59)
       Weight: 1.7–2.3 kg (2.0); 3.8–5.1 lb. (4.4)

BV01. **Black Vulture.** Adults and juveniles are almost alike. Appear all black except for large whitish primary panels and whitish legs. Note legs almost reach tip of short tail and wings held in shallow dihedral. This bird is about a year old and shows retained juvenile secondaries and smooth face. [TX, Oct]

BV02. **Black Vulture sunning.** Uppersides appear all black except for large whitish primary panels; compare these panels to pale areas on Turkey Vulture upperwings in photo TV07. Sunning posture is common. This bird is a juvenile. [FL, Aug]

BV03. **Adult Black Vulture.** Plumage appears black overall and is somewhat iridescent. Dark grayish head skin is wrinkled, outer half of beak is yellowish or ivory, and legs are whitish. Compare head and beak to Turkey Vultures in photos TV04–07. [FL, Aug]

BV04. **Juvenile Black Vulture.** Differs from adult by smooth blackish head skin that is covered by black fuzzy down, by dark beak, and by body plumage that is less iridescent. [FL, Aug]

CALIFORNIA CONDOR *(Gymnogyps californianus)*

Photos CC01–03

The California Condor is a huge black raptor; the largest in North America. They exhibit a variety of subtly different plumages as they progress from juvenile to adult; the main differences are in head coloration and amount of mottling in white areas of upperwings and underwings. Sexes are alike in plumage and size. **Adults** are overall black, except for large white areas on the underwing coverts and a silvery cast to uppersides of secondaries and greater coverts. Head and swollen neck are orange, and legs are whitish. **Juveniles** are similar in plumage but their heads and necks are narrower and dusky, and underwing coverts are mottled black and white. **Immatures** gradually get more white on underwing coverts and heads become thicker and more orangish.

Best field marks are huge size, very broad wings with deeply slotted primaries and bold white triangles on undersides, and orange to orange-yellow puffy heads of all but juveniles, which are blackish. Wings are held level when soaring and gliding but may be held in shallow dihedral under some soaring conditions.

Similar Species:
Turkey Vultures (photos TV01–07) are much smaller, have much longer tails, and fly with wings typically in a strong dihedral.
Golden Eagles (photos GE01–10) are noticeably smaller with longer tails. All show golden hackles on crown and nape.
Bald Eagle (photos BE2–10, 13–19) non-adults can also show white underwing coverts, but are noticeably smaller with longer tails.
Range: Some captive raised juveniles have been released back into the former range in southern California, with more releases planned there, in the Grand Canyon, and in New Mexico.
Measurements:
> **Length:** 109–127 cm (117); 43–50 in. (46)
> **Wingspan:** 249–300 cm (278); 98–118 in. (109)
> **Weight:** 8.2–14.1 kg (10.5); 18–31 lb. (23)

CC01. **Adult California Condor.** Overall black with white triangles on underwings. Note orangish head, whitish legs, and short tail. Underwing coverts appear somewhat mottled due to molt. [CA, Sept]

CC02. **Adult California Condor.** Huge size and orangish puffy head and neck are distinctive. Note reddish bare area on breast and whitish legs. [CA, Sept]

CC03. **Basic I California Condor.** Juveniles and Basic I birds have blackish heads and reddish-brown eyes [CA, June]

OSPREY *(Pandion haliaetus)* Photos O01–06

The Osprey, a large, long-winged whitish raptor, has two recognizably different plumages: adult and juvenile. Sexes of adults are similar in plumage, but females are noticeably larger and have, on average, heavier dark breast markings that often form a necklace (there is complete overlap in this character, so it is not possible to sex them by it). **Adults** have white heads with black eye-lines, white underparts and underwing coverts, dark carpal patches on underwings, and dark brown backs and upperwing coverts. Adults have yellow eyes. **Juveniles** are similar to adults but have pale feather edges on back and upperwing coverts giving a scaly appearance to upperparts, short black streaks on white areas of crown and nape, rufous wash on upper breasts (which fades quickly) and underwing coverts, whitish mottling on black carpal patches, and wider white band on tips of tail and secondaries. Undersides of juveniles' secondaries are paler than those of adults. Juveniles have red to orange-yellow eyes.

Best field marks are gull-like crooked wings of flying birds, dark eye-line and lack of supra-orbital ridge on the white head, white underparts, and black carpal patches on underwings.

Similar Species:
Bald Eagle non-adults (photos BE14–15, 18) can appear somewhat similar but are larger, with dark breasts in all plumages.
Large gulls can appear similar but are larger headed, have pointed wingtips, and lack dark carpal patches on underwings.
Range: Usually found near water. Throughout North America on migration and many areas in summer. Breeds: Gulf Coast, Florida, Atlantic Coast, Great Lakes, Boreal Forest from e. Canada to w. Alaska, Pacific Northwest forests, Coastal California, and Rocky Mountains. Sporadic elsewhere. Most of population migrates south for winter, but a few birds winter along Gulf coast, in southern Florida, and in central and southern California.
Measurements:
       Length: 51–62 cm (56); 20–25 in. (23)
   Wingspan: 149–171 cm (162); 59–67 in. (63)
     Weight: 1.0–1.9 kg (1.6); 2.2–4.2 lb. (3.5)

O01. **Adult Osprey.** Note white head with dark eye-line, dark carpal patch on underwing, crooked wing, and lack of supra-orbital ridge. [FL, Jan]

O02. **Adult Osprey.** Note dark eye-line, long wings, and underwing and tail patterns. [TX, Oct]

O03. **Juvenile Osprey.** Note buffy wash on wing linings and whitish mottling on carpal patches. [NJ, Oct]

O04. **Juvenile Osprey.** Juveniles have pale feather edges on back and upperwing coverts; this produces a scaly appearance to upperparts. White areas of crown and nape show short black streaking. Note wide white band on tail tip. [CT, Sept]

O05. **Adult Osprey.** Backs and upperwing coverts of adults do not have extensive pale feather edges. Raised hackles (nape féathers) result in large-headed appearance. Wingtips extend beyond tail tip. Dark "necklace" indicates that this is probably an adult female. (Photo is of an adult female.) [FL, Jan]

O06. **Adult Osprey.** Lack of supra-orbital ridge makes head appear pigeon-like. Lack of breast markings indicates that this is most likely (but not certainly) an adult male. Note also the long legs and white stripe between folded wings and back. [FL, Aug]

Kites

Kites are medium sized raptors occurring in subtropical into temperate areas. The five species that occur in North America are rather specialized feeders.

For field identification purposes, they can be placed into two types by wing shape: pointed-winged kites and paddle-winged kites.

The more widespread species, the Swallow-tailed, White-tailed, and Mississippi Kites, all have pointed wings. The limited-range species, the Snail and Hook-billed Kites, have paddle-shaped wings.

HOOK-BILLED KITE   *(Chondrohierax unicinatus)*

Photos HB01–06

The Hook-billed Kite has three recognizably different plumages: adult male, adult female, and juvenile. Sexes are similar in size. Adults have white eyes and two wide pale bands in dark tails (gray above, white below), but one band is often covered below by undertail coverts and above by primaries on perched kites. **Adult males** are overall gray, with white barring on belly. **Adult females** have brown backs, distinctive wide rufous collars, rufous and white barred underparts, and a rufous cast to inner primaries. **Juveniles** are brown on backs like adult females but have whitish underparts that show a variable amount of dark brown barring from heavy to none. They also have three narrow light tail bands, light brown eyes, and lack rufous on inner primaries.

Best field marks are large hooked beaks and paddle-shaped wings that pinch in to body on trailing edge of wings. Up close a unique small greenish bare patch of skin over and in front of the eye and on cere and the lack of a supra-orbital ridge (giving pigeon-headed look) are noticeable. This species has a dark color morph that occurs in much of its range but not in Texas.

Similar Species:
Gray Hawk adults (photos GH01, 03, 05) are also gray and appear similar to adult male Hook-bills but have dark eyes, more pointed wingtips, more finely barred underparts, pale underwings, and much smaller beaks.
Red-shouldered Hawk adults (photos RS01–03) can appear similar in flight to adult female Hook-bills but have narrower wings that show white crescent-shaped panels near tips of underwings and more and narrower white tail bands.
Harris' Hawks (photos HH01–05) also have rather paddle-shaped wings and can appear similar in flight but have dark bodies and white tail coverts and base and tip of tail.
Roadside Hawks (photos RH01–03) also have pale eyes and barred underparts but eye color is yellow or yellow-orange, not white, and they lack large hooked beaks. Adult female Hook-bills in Texas show rufous on inner primaries, but n. Mexican Roadside Hawks usually do not.
Range: Uncommon and local resident along the Rio Grande River in the lower Rio Grande valley of Texas from Falcon Dam to Brownsville.
Measurements:
      Length: 43–51 cm (46); 16–20 in. (18)
  Wingspan: 87–98 cm (92); 34–38 in. (36)
     Weight: 215–353 g (277); 8–12 oz (10)

HB01. **Adult male Hook-billed Kite.** Overall dark gray except for barred belly and whitish spotting on primaries. Long tail shows two wide white bands. Note paddle-shaped wings that pinch in to body on trailing edges. [TX, May]

HB02. **Adult female Hook-billed Kite.** Overall brownish with rufous barred underparts, pale flight feathers with heavy dark barring, and paddle-shaped wings. Note rufous on inner primaries. Long tail shows two wide white bands. [TX, May]

HB03. **Juvenile Hook-billed Kite.** Like adult female except underparts have dark barring and inner primaries lack rufous. Note pale brown eye, tail with three narrow light bands, and paddle-shaped wings. Some have little or no barring on underparts. [TX, Oct]

HB04. **Adult male Hook-billed Kite.** Overall dark gray but with whitish barring on belly. Note large beak, white eye, bare green face skin, short legs. Lack of supra-orbital ridge results in pigeon-headed appearance. [TX, May]

HB05. **Adult female Hook-billed Kite.** Underparts are barred rufous and white. Note large beak, white eye, bare green face skin, and lack of supra-orbital ridge giving pigeon-headed appearance. [TX, Feb]

HB06. **Adult female Hook-billed Kite.** Rufous collar contrasts with brown back and dark crown. Note large beak, white eye, bare face skin. [TX, May]

SWALLOW-TAILED KITE   (*Elanoides forficatus*)

Photos SW01–04

The Swallow-tailed Kite is distinctive and unlikely to be mistaken for any other raptor. Sexes are similar in size and identical in plumage; plumage of juvenile is almost like that of adult. **Adults** are white on head, body, and underwing coverts and black on back, tail, flight feathers, and upperwing coverts. Upperparts are somewhat two toned: lower back, flight and tail feathers, and primary and greater secondary coverts have a grayish cast that contrasts somewhat with black of upper back and lesser and median secondary upperwing coverts. Upperparts also show purplish iridescence. Adult eyes are red. **Juveniles** are like adults except for brown eyes, shorter tails, and white tips to flight and tail feathers. Their upperparts show a greenish iridescence.

Best field marks are long swallow-like deeply forked tail and bold black and white plumage. Hunting is almost always on the wing; feeding is often on the wing.

Similar Species: No other raptor is similar.
Range: Local in Florida and along the Gulf coast west to e. Texas and the Atlantic coast north to South Carolina. In spring and early fall, individuals wander from breeding area far to the west and north. Entire population migrates into South America for the winter.
Measurements:
　　　　Length: 52–62 cm (58); 20–25 in. (22)
　　Wingspan: 119–136 cm (130); 47–54 in. (51)
　　　Weight: 325–500 g (430); 11–18 oz (15)

SW01. **Adult Swallow-tailed Kite.** Note deeply forked long black tail, pointed wingtips, white body and head, and black and white underwing pattern. They often feed on the wing. [FL, Aug]

SW02. **Juvenile Swallow-tailed Kite.** Like adult in SW01 but with shorter, less deeply forked tail. [FL, Aug]

SW03. **Swallow-tailed Kite.** Black area across upper back and lesser upperwing coverts contrasts with rest of grayish upperparts. Lack of supra-orbital ridge gives head pigeon-like appearance. [FL, Aug]

SW04. **Adult Swallow-tailed Kite.** Lack of supra-orbital ridge gives head pigeon-like appearance. Note forked tail and overall black and white appearance. Eye color difficult to see in the field. [FL, Aug]

WHITE-TAILED KITE (*Elanus leucurus*) Photos WK01–06

The White-tailed Kite is falcon-shaped and has two recognizable plumages: adult and juvenile. Sexes are similar in size. **Adults** have white heads, tails, and underparts. Crown, back, some upperwing coverts, uppersides of flight feathers, and central tail feathers are gray. Lesser and median upperwing coverts are black and form black shoulder on perched kites. Underwings show white coverts and secondaries, black carpal patches, and grayish-black primaries. Eyes are scarlet. **Juveniles** are similar but back and crown are brownish gray with wide white feather edges, breast has rufous wash, flight feathers and greater and median upperwing coverts have white tips, and white tail shows a narrow gray subterminal band. Eyes are orangish-yellow. Juveniles undergo a post-juvenile molt shortly after becoming independent and become much more adult-like. However, they retain flight and a variable number of tail feathers and some white-tipped greater secondary and all primary upperwing coverts. Eyes gradually become more orangish and later reddish.

Best field marks are combination of falcon shape, white tail, and black carpal patches on underwings of flying kites and black shoulder patches of perched kites.

Note: The AOU in 1983 combined this species with the Black-shouldered Kite (*Elanus caeruleus*) of Africa, southern Europe, and Asia, but recent studies show that it is quite distinct. The AOU in 1994 has again recognized the White-tailed Kite as a separate species, *Elanus leucurus*.

Similar Species:
Mississippi Kites (photos MK01–12) are similar in shape and size but have dark tails and bodies in all plumages.
Northern Harrier adult males (photos NH01, 05, 08) are similar in shape and coloration but lack black carpal patches and have all gray heads, whitish uppertail coverts that form noticeable white patches, and black terminal band on trailing edge of underwings.
Range: Western California and southwestern Oregon and locally in southeastern Arizona, Gulf coast from Texas to Florida, and peninsular Florida.
Measurements:
       Length: 36–41 cm (38); 14–16 in. (15)
  Wingspan: 99–102 cm (101); 37–40 in. (39)
     Weight: 305–361 g (330); 10–13 oz (11.6)

WK01. **Adult White-tailed Kite.** Note falcon shape, white head, body, and tail, grayish-black primaries, and black carpal patches on underwings. [TX, Oct]

WK02. **Older juvenile White-tailed Kite.** Like adult in photo WK01 but with narrow gray subterminal tail band. [TX, Oct]

WK03. **Fledgling White-tailed Kite.** Like adult in photo WK01 but with rufous wash across breast, brownish-gray crown, and narrow gray subterminal tail band. Note also white tips on flight feathers. Fledglings have more rounded wingtips; outer primaries are not fully grown. [CA, July]

WK04. **Adult White-tailed Kite.** Note white head, red eye, and black shoulder patch. Wingtips do not reach tail tip. [TX, Dec]

WK05. **Older juvenile White-tailed Kite.** Note gray streaks on crown, orangish-yellow eye, and white tips of back feathers, upperwing coverts, and flight feathers. Wingtips do not reach tail tip. [TX, Dec]

WK06. **Fledgling White-tailed Kite.** Note rufous wash across breast, orangish-yellow eye, and scalloped appearance of back and crown. [CA, July]

SNAIL KITE   *(Rostrhamus sociabilis)*   Photos SN01-07

The Snail Kite has three recognizably different plumages: adult male, adult female, and juvenile. All show white tail coverts and dark tail with wide white base and tip. Sexes are similar in size. Adults have red eyes and orange legs. **Adult male** is mostly dark gray. Black uppersides of flight feathers contrast with gray wing coverts. **Adult female** is mostly dark brown, except for some white head markings, tawny feather edges on upperparts, white streaking on underparts, and dark narrow barring on grayish undersides of flight feathers. **Juveniles** are similar to adult females but have tawny head markings, more extensive tawny edges on upperparts, and tawny underparts with dark streaking. Juveniles have brown eyes and dull yellow cere and legs.

Best field marks are thin hooked beak, square tail with white base and tip, and arched or cupped, paddle-shaped wings of flying kites.

Similar Species:
Northern Harriers (photos NH05, 06) also show white at the base of tail but not on the tail. They fly with their narrower wings in a dihedral.
Range: Found in freshwater marshes in central to southern Florida.
Measurements:
      Length: 41–47 cm (44); 16–19 in. (17)
  Wingspan: 104–112 cm (108); 41–44 in. (42)
     Weight: 340–520 g (427); 12–21 oz (15)

SN01. **Adult male Snail Kite.** Overall dark gray with whitish spotting on underwings (lacking on older males) and white tail coverts and base of tail. Note red eye; bright orange cere, face skin, and legs; and long thin hooked beak. Snails are carried either in foot or beak. [FL, Aug]

SN02. **Adult female Snail Kite.** Overall dark brown with white streaking on dark underparts, heavily barred flight feathers, white undertail coverts and base and tip of tail, and paddle-shaped wings. [FL, Aug]

SN03. **Juvenile Snail Kite.** Similar to adult female (photo SN02) but with dark streaking on tawny underparts and more extensive tawny markings on head. Secondaries are usually darker than primaries. [FL, Aug]

SN04. **Adult male Snail Kite.** Note two-toned pattern on upperparts, white uppertail coverts and base and tip of tail, and arched or cupped wings that are somewhat paddle-shaped. [FL, Aug]

SN05. **Adult male Snail Kite.** Overall dark gray, with red eye; bright orange cere, face skin, and legs; and thin hooked beak. [FL, Aug]

SN06. **Adult female Snail Kite.** Dark brown overall, with white areas above and below red eye, thin hooked beak, yellow cere and face skin, and orangish legs. Females often show some tawny tips on back feathers and upperwing coverts. [FL, Aug]

SN07. **Fledgling Snail Kite.** Similar to adult female but has brown eye, yellow cere and legs, pale face skin, more extensive tawny markings on head and upperparts, and tawny, heavily streaked underparts. [FL, Aug]

MISSISSIPPI KITE (*Ictinia mississippiensis*) Photos MK01–12

The Mississippi Kite, a dark falcon-shaped kite, has three recognizably different plumages: adult, subadult, and juvenile. Sexes are similar in size and adults have slightly different plumages. Adults are overall dark gray, except for black primaries and tail and white uppersides of secondaries. Their eyes are red and ceres are gray. **Adult male** has whitish head and some rufous on inner primaries. **Adult female** has darker head, less rufous in primaries, and paler undersides of primaries and tail, latter with dusky band on tip. Undertail coverts of female are usually barred with white. **Juveniles** have brownish-gray head with obvious short buffy superciliary lines and darker back and upperwing coverts, both with whitish feather edges. Their white underparts are heavily streaked with dark rufous-brown and flight feathers are dark brown; secondaries have narrow white tips. Dark brown tail usually shows two or three narrow white bands. Eyes are brown, and cere is yellow. **Subadults** are juveniles returning in spring with adult-like gray body and some wing coverts but retained flight and tail feathers. They become more adult-like by molt through summer. When they again migrate in fall, only the juvenile secondaries, some underwing coverts, and some tail feathers have not been replaced. Gray body usually shows some retained juvenile feathers and often some whitish spots. Eyes and cere like those of adult.

Best field marks are overall gray color and falcon shape, short outer primary, dark spot in front of eye, and light, graceful flight.

Similar Species:
White-tailed Kites (photos WT01–06) are similar in silhouette but are whitish overall with black carpal patches on underwings and black shoulders on perched birds.
Peregrines (photos P01–13) may appear similar in overall shape but are larger, have dark mustache marks, and fly much faster and more purposefully.
Range: Summer breeder from North Carolina and n. Florida west to Kansas, southeastern Colorado, and Texas and north along Mississippi River to s. Illinois and Indiana, and west locally to Arizona. In spring, individuals, usually subadults, wander far afield to west and north.

Measurements:

Length:	31–37 cm (35);	12–15 in. (14)
Wingspan:	75–83 cm (78);	29–33 in. (31)
Weight:	240–372 g (278);	8–13 oz (10)

MK01. **Adult male Mississippi Kite.** Gray body and underwing coverts and pale gray head. Note short outer primary, black flared, square-cornered tail, and narrow pale band on trailing edge of inner .wing. [CO, June]

MK02. **Adult female Mississippi Kite.** Similar to adult male (photo MK01) but head grayer, undertail coverts barred whitish, and underside of tail paler with darker terminal band. [CO, June]

MK03. **Subadult Mississippi Kite.** Early summer birds have retained juvenile tail, flight feathers, and underwing coverts, as well as some body feathers. [CO, June]

MK04. **Subadult Mississippi Kite.** Late summer birds are more like adults and are actively molting, with many new adult primaries and tail feathers and mostly gray underwing coverts. All secondaries are retained from juvenile plumage. [CO, Aug]

MK05. **Adult Mississippi Kite.** Note white secondaries and rufous in primaries. Male shown; females have less rufous in primaries. [CO, June]

MK06. **Subadult Mississippi Kite.** Secondaries are dark, not white as are those of adults. Note retained juvenile tail. [CO, Aug]

MK07. **Juvenile Mississippi Kite.** Note heavily streaked underparts, narrow white bands in tail, and narrow pale terminal band on inner wing. Outer primaries usually show whitish areas on undersides (see photo MK03). A few lack tail banding. [OK, Aug]

MK08. **Adult male Mississippi Kite.** Note whitish head, whitish bar across wing and solid black tail with notched tip. Black spot in front of red eye makes eye appear larger. Wingtips extend beyond tail tip. [OK, Aug]

MK09. **Adult female Mississippi Kite.** Compared to adult male, has darker head, whitish barring on undertail coverts, and pale underside of tail showing darker band on tip. [CO, July]

MK10. **Subadult Mississippi Kite.** Mostly gray underparts have a few retained juvenile feathers and some new feathers with white spots. Eye is red. Wingtips extend beyond tip of retained juvenile tail. [CO, June]

MK11. **Subadult Mississippi Kite.** Appears much like adult, but retained juvenile secondaries are dark and retained juvenile tail has narrow pale bands. [CO, June]

MK12. **Juvenile Mississippi Kite.** Note brown eye, short pale superciliary line, black spot in front of eye, pale edges of back and upperwing coverts, and whitish tips of primaries. Wingtips extend to tail tip. [OK, Aug]

NORTHERN HARRIER (*Circus cyaneus*)   Photos NH01–10

The Northern Harrier, the only harrier in North America, has three recognizably different plumages: adult male, adult female, and juvenile. All show a facial disk and a white patch on uppertail coverts. Females are separably larger than males. Adults have yellow eyes, except for younger females, which may have dark to light brown eyes. **Adult males** are darkish gray on head, back, breast, upperwing coverts, and uppertail. Their white breasts have rufous spots. **Adult females** have brownish heads, backs, and upperwing coverts. Their buffy underparts are heavily streaked dark brown. Dark secondaries on underwings show two wide whitish bands. **Juveniles** are similar to adult females but have rufous underparts (which fade to creamy by spring) with dark streaking, if present, restricted to upper breast and sides. They show dark patches on underwings composed of dark secondaries, axillaries, and greater and median secondary coverts. Eye colors of juveniles are dark brown on females and pale gray-brown on males.

   Best field marks are white patch on uppertail coverts, facial disk, and dark hood in all plumages, and slow quartering flight close over the ground with wings held in a strong dihedral.

Similar Species:
Northern Harriers are usually confused with other raptors only when soaring or gliding at considerable altitude or when perched uncharacteristically in a tree.
Rough-legged Hawk light morph (photos RL04–05,09) shows white at base of tail and has black carpal patches on underwing.
Turkey Vulture (photo TV02) also flies with wings in a strong dihedral but is blackish brown overall and lacks white uppertail coverts.
Swainson's Hawk (photos SH01–13) also flies with wings in a strong dihedral and has white on uppertail coverts but has two-toned underwings and pointed wingtips.
Red-shouldered Hawk adults (photos RS01, 02) can appear similar in flight to high flying juvenile or adult female harriers but have crescent-shaped white primary panels and lack dark secondary panels on underwings.
Peregrines (photos P01–05) and harriers can appear quite similar when gliding or stooping, but Peregrines lack white uppertail coverts and dark hoods.
Range: Breeds in open habitats throughout n. and c. U.S. and Canada; moves south into s. U.S. and farther south in winter.
Measurements:
    Length: Male 41–45 cm (43); 16–18 in. (17)
              Female 45–50 cm (48); 18–29 in. (19)
    Wingspan: Male 97–109 cm (103); 38–43 in. (41)
              Female 111–122 cm (116); 43–48 in. (46)
    Weight: Male 290–390 g (346); 10–14 oz (12)
              Female 390–600 g (496); 14–21 oz (18)

NH01. **Adult male Northern Harrier.** Appears overall white below with dark gray head and neck giving hooded look and black tips on primaries and secondaries; the latter form a black band on trailing edges of wings. Note long tail and fine rufous streaking on underparts. Adult eye is lemon yellow. [CO, May]

NH02. **Adult female Northern Harrier.** Dark brown head and neck give hooded look. Note darkly streaked buffy underparts and darkish patches on underwings. Adult female has noticeable white bands through secondaries. Long tail shows even-width dark and light bands when spread. [AZ, Feb]

NH03. **Juvenile Northern Harrier.** Similar to adult female but dark streaking on rufous underparts restricted to upper breast and sides and secondary patches appear darker. Note facial disk and dark brown head and neck giving hooded look. [NJ, Oct]

NH04. **Juvenile Northern Harrier.** By spring, color of juveniles' underparts has faded to creamy. Note darkish patch on underwing composed of greater secondary coverts, axillaries, and secondaries and dark brown head and neck giving hooded look. [CO, Apr]

NH05. **Adult male Northern Harrier.** Note gray upperparts, and white uppertail coverts. [CO, May]
NH06. **Juvenile Northern Harrier.** White patch on uppertail coverts is diagnostic. Tawny bars on upperwings are shared by adult females. [CT, Oct]

NH07. **Juvenile Northern Harrier.** Shows wings held in the classic dihedral. All harriers show distinct facial disk. Note dark brown eye of juvenile female and streaking restricted to flanks. [TX, Nov]

NH08. **Adult male Northern Harrier.** Dark gray head and upperparts. Note lemon yellow eye, facial disk, and long tail. [CA, Nov]

NH09. **Juvenile Northern Harrier.** Juveniles have relatively unstreaked rufous underparts. All harriers show distinct facial disk. Note dark brown eye of juvenile female. [TX, Nov]

NH10. **Adult and juvenile female Northern Harriers.** Adult female (right) has heavily streaked buffy underparts and yellow eye. Juvenile female (left) has unstreaked rufous underparts and dark brown eyes. Both show facial disks and white uppertail covert patches. [CA, Jan]

Accipiters

Accipiters are short-winged, long-tailed forest raptors. Three species occur in North America: Northern Goshawk, Cooper's Hawk, and Sharp-shinned Hawk, in order by decreasing size. They are all characterized by strongly barred flight feathers and three or four pairs of even-width dark and pale brown tail bands.

There is no size overlap between species, nor even between sexes within species but, as size is difficult to judge in the field, plumage similarities make field identification a challenge, particularly that of separating Cooper's and Sharp-shinned Hawks.

SHARP-SHINNED HAWK   (*Accipiter striatus*)
Photos SS01–07

The Sharp-shinned Hawk, our smallest *Accipiter*, has two recognizably different plumages: adult and juvenile. Females are separably larger than males. **Adults** have blue-gray upperparts and rufous and white barred underparts. The crown is same color as back without sharp line of contrast on nape. Adult eyes progress from orange to red with age. **Juveniles** have brown backs and upperwing coverts and creamy underparts that are usually heavily streaked with reddish-brown on both belly and breast, with heavy barring on flanks. However, some juveniles, especially males, have underparts with narrow dark brown streaking. Backs and upperwing coverts usually show a few white spots. Juveniles have pale superciliaries and yellow eyes.

Best field marks are small rounded head with eye centrally placed, stick-like legs, and square-tipped tail (all tail feathers are about the same length), with at most a narrow white band on tip. In flight, the head barely projects beyond wrists, which are usually thrust forward.

Similar Species:
Cooper's Hawks (photos C01–06) are always separably larger and more robust, but often appear almost identical in field. Perched Cooper's usually show larger squarish head (hackles raised, but can show rounded head when not raised), stouter legs, and longer tail, usually with wide white terminal band. In flight, their heads project way beyond wrists and long tails show rounded tip and shorter outer feathers, often with wide white bands noticeable. Wrists are usually not thrust forward much, thus leading edge of wing is straight and perpendicular to body. Adult Cooper's have crown darker than back, with noticeable line of contrast with paler nape. Juvenile Cooper's usually lack pale superciliary.
Merlins (photos M09–10), particularly Taiga adult females and juveniles, appear dark brown like juvenile Sharpies when perched, but their tails are dark with narrow pale bands, their eyes are dark, their legs are short, and their heads look squarish, not roundish.
Range: Breeds in forests of eastern, northern, and mountainous western N. America. Concentrates on migration in a few locations on Great Lakes and Atlantic coast. Winters throughout U.S. except northern Great Plains, as well as farther south.
Measurements:
      Length: Male 24–27 cm (26); 9–11 in. (10)
                Female 29–34 cm (31); 11–13 in. (12)
  Wingspan: Male 53–56 cm (54); 20–22 in. (21)
                Female 58–65 cm (62); 23–26 in. (25)
     Weight: Male 87–114 g (101); 3–4 oz (3.6)
                Female 150–218 g (177); 5–8 oz (6)

SS01. **Adult Sharp-shinned Hawk.** Note short neck and head and short, square-tipped tail with thin white terminal band. Compare to adult Cooper's Hawk (photos C01–02). [MN, Sept]

SS02. **Juvenile Sharp-shinned Hawk.** Note short neck and head, streaking on underparts extends onto belly, and short square-tipped tail. Compare to juvenile Cooper's Hawk (photo C03). [MN, Sept]

SS03. **Juvenile Sharp-shinned Hawk.** Head barely projects beyond wrists in glide. Note square-tipped tail with straight narrow white terminal band. [NJ, Oct]

SS04. **Adult Sharp-shinned Hawk.** Small rounded head has large orange to red eye. Note stick-like legs. Outer tail feathers are same length as others. Compare to adult Cooper's Hawk (photo C05). [CA, Jan]

SS05. **Adult Sharp-shinned Hawk.** Crown is same color as back, without line of contrast. Tail is square-tipped, with a narrow white terminal band. Compare to adult Cooper's Hawk (photo C05). [CA, Jan]

SS06. **Juvenile Sharp-shinned Hawk.** Small rounded head has bright yellow eye. Streaking of underparts extends onto belly. Legs are stick-like. Outer tail feathers are same length as others. Compare to juvenile Cooper's Hawk (photo C06). [NJ, Oct]

SS07. **Juvenile Sharp-shinned Hawk.** Note pale superciliary. White tail tip is narrow. Notice that tail tip appears a bit rounded, a feature shown by many female Sharpies. Compare to juvenile Cooper's Hawk (photo C06). [CT, Sept]

COOPER'S HAWK   *(Accipiter cooperii)*   Photos C01-07

The Cooper's Hawk, our middle-sized *Accipiter*, has two recognizably different plumages: adult and juvenile. Females are separably larger than males. **Adults** have blue-gray upperparts and rufous and white barred underparts. Their blackish crowns are darker than their backs and show a sharp line of contrast with paler nape. Adult eyes progress from orange to red with age. **Juveniles** have brown backs and upperwing coverts and creamy to white underparts that are narrowly streaked dark brown. Backs and upperwing coverts show many white spots and wide rufous feather edges. Juveniles often have a tawny cast to head and neck. Juveniles' eyes are straw colored

Best field marks are large square head when hackles are raised, then showing eye closer to beak than nape, stout legs, and rounded tail (outer tail feathers are noticeably shorter than central ones) with a wide white band on tip, often worn off by spring. In a soar, wings are held in a dihedral and head projects far beyond wrists, which are usually not thrust forward, resulting in straight leading edges of wings; in a glide, head usually projects beyond wrists, which are somewhat thrust forward.

Similar Species:
Sharp-shinned Hawks (photos SS01–07) are always separably smaller, but often appear almost identical in field. See under that species for distinctions.
Northern Goshawk juveniles (photos G04–05, 07) are separably larger, but can appear similar in the field. They have wide pale superciliaries, heavily streaked bellies and underwing and undertail coverts, and paler, more heavily marked backs and upperwing coverts. Also, their tail banding is irregular, with white highlights between bands. In flight, Goshawks usually show a narrow tawny diagonal band on each upperwing and have longer, more tapered wings.
Range: Breeds in open forests over most of the U.S. and southern Canada, except Florida peninsula and northern Great Plains. Winters throughout the U.S. except for n. Great Plains.
Measurements:
> **Length:** Male 37–41 cm (39); 14–16 in. (15)
> Female 42–47 cm (45); 16–19 in. (18)
> **Wingspan:** Male 70–77 cm (73); 28–30 in. (29)
> Female 79–87 cm (84); 31–34 in. (33)
> **Weight:** Male 302–402 g (341); 10–14 oz (12)
> Female 479–678 g (528); 17–24 oz (19)

C01. **Adult Cooper's Hawk.** Soars with leading edge of wing straight (often with wings held in dihedral). Note long tail with shorter outer tail feathers and wide white tip. Compare to adult Sharp-shinned Hawk (photo SS01). [NJ, Oct]

C02. **Adult Cooper's Hawk.** Glides with wrists forward; head projects far beyond wrists. Outer tail feathers are shorter than others. Long tail has rounded tip. Compare to adult Sharp-shinned Hawk (photo SS01). [CO, Apr]

C03. **Juvenile Cooper's Hawk.** Head and tail are longer than those of Sharpies. Belly streaking is sparse or absent. Note long tail with shorter outer tail feathers and wide white tip. Compare to Sharp-shinned Hawk (photo SS02). [NJ, Oct]

C04. **Juvenile Cooper's Hawk.** Head projects far beyond wrists on gliding birds. Note rounded corners of tail tip. Compare to juvenile Sharpie in photo SS03. [NJ. Oct]

C05. **Adult Cooper's Hawk.** Crown is darker than back, with sharp line of contrast on nape. When hackles are raised, rear of head appears square and eye closer to beak. Outer tail feathers are shorter than others on long tail. Compare to adult Sharp-shinned Hawk (photos SS04-05). Cooper's often sit on poles in open, particularly in the West. [TX, Dec]

C06. **Juvenile Cooper's Hawk.** Head shows square nape (hackles raised) and eye closer to beak than to nape. Fine dark brown streaking is sparse on belly. Legs are stout, not stick-like. Tail tip appears rounded. Compare to juvenile Sharp-shinned Hawk (photo SS06). [NJ, Oct]

C07. **Juvenile Cooper's Hawk.** Head has a tawny cast and narrow pale superciliary. Back and upperwing coverts show many white spots. Long tail has wide white tip. Compare to juvenile Sharp-shinned Hawk (photo SS07) and Goshawk (photo G07). [NJ, Oct]

NORTHERN GOSHAWK   (*Accipiter gentilis*)
Photos G01–07

The Northern Goshawk, our largest *Accipiter*, has two recognizably different plumages: adult and juvenile. Females are noticeably larger than males. **Adults** are distinctive: blackish head with wide white superciliaries, finely barred (some young females are coarsely barred) gray underparts, white unmarked undertail coverts, and blue-gray backs and upperwing coverts. Upperwings are two-toned: dark flight feathers contrast with paler coverts. Adult eyes vary from orange to deep red to mahogany with age. **Juveniles** have heavily mottled brown back and upperwing coverts, heavily streaked belly and undertail coverts, and irregular tail bands with white highlights between bands. Juveniles' eyes are yellow.

Best field marks are long wings that are tapered in soar and glide and appear pointed when flapping and wedge-shaped tail tip. Adult's dark head with white superciliary lines and juvenile's irregular tail banding and tawny bar on upperwing are also good field marks.

Similar Species:
Cooper's Hawk juveniles (photos C01–06) are separably smaller, but can appear similar in field. See under that species for distinctions.
Gyrfalcon adults (photos GY01, 05) can appear similar to adult Goshawks—both show pointed wingtips in flight—but usually have two-toned underwings, uniform upperwings (adult Gos have two-toned upperwings), and dark wingtips and lack dark hood with wide superciliaries. Wingtips extend over half way down tail on perched Gyrs.
Red-shouldered Hawk juveniles (photos RS12–17) can appear similar. See under that species for distinctions.
Broad-winged Hawk juveniles (photos BW09–11) can appear similar. See under that species for distinctions.
Range: Breeds in forests of northern U.S. south in Appalachian Mts. to West Virginia, throughout Canada and Alaska, and in forested Western mountains. Juveniles move south in winter; adults also during periodic low cycles of prey, occurring every nine to eleven years.
Measurements:
  Length: Male 46–51 cm (49); 18–20 in. (19)
          Female 53–62 cm (58); 21–24 in. (23)
  Wingspan: Male 98–104 cm (101); 38–41 in. (39)
          Female 105–115 cm (108); 41–45 in. (43)
  Weight: Male 677–1014 g (816); 24–36 oz (29)
          Female 758–1214 g (1059); 26–43 oz (37)

G01. **Adult male Goshawk.** Adults have deep red eyes and vermiculated underparts, with barring finer on males. Note unmarked secondaries and white undertail coverts. [MN, Oct]

G02. **Adult female Goshawk.** Adults have orange to deep red eyes and vermiculated underparts, with barring coarser on females. Note wide pale superciliary on dark head, unmarked secondaries, and white undertail coverts. [MN, Oct]

GO3. **Young adult female Goshawk.** First plumage adults have yet coarser barring on underparts and more distinct dark bands on undertails. Note retained juvenile secondaries and tail feathers. [MN, Oct]

G04. **Juvenile Goshawk.** Undersides of long, tapered wings show heavily barred flight feathers and heavily marked coverts. Underparts, including undertail coverts, are heavily marked. Wedge-shaped tail tip is distinctive. [MN, Oct]

G05. **Juvenile Goshawk from above.** Tawny bar across upperwings, pointed wingtips, and irregular tail banding are distinctive. [MN, Oct]

G06. **Adult Goshawk.** Dark head has white superciliary, gray underparts are finely barred, and fluffy undertail coverts are white. [CT, Apr]

G07. **Juvenile Goshawk.** Wide pale superciliary, tawny bar on upperwing coverts, dark spots on undertail coverts, long primary projection, and irregular bands and white highlights on upper tail are distinctive. Underparts are heavily streaked. [NJ, Nov]

Buteonines

The buzzards of the genus *Buteo* and closely related genera are all characterized by robust bodies, long, broad wings, and short to long tails. All soar regularly, and many hover. Wing and tail patterns, wing shape, and sometimes behavior, help in their field identification. Many species have a dark (melanistic) or a rufous (erythrystic) color morph or both in addition to the normal or light morph, and two occur only in the dark morph.

Ten species of *Buteo* and two closely related species occur in our area. Six buteos are widespread throughout eastern or western North America or both; the other four and the two others breed on our southern periphery.

As in most raptor species, the tails of juvenile buteonines are noticeably longer than those of adults, but unlike other species, the wings are narrower than those of adults in most species.

Light-morph birds of most species show a dark comma at the wrist on each underwing.

COMMON BLACK HAWK   (*Buteogallus anthracinus*)
Photos BH01–07

The Common Black Hawk is a large dark buteonine that has two recognizably different plumages: adult and juvenile. Females are somewhat larger than males. Wingtips of perched birds fall somewhat short of tail tip. **Adults** are coal black overall except for single wide white tail band. Orangish-yellow face skin and base of beak are distinctive. Very wide wings of flying adults show small whitish commas at base of outer primaries and faintly marked secondaries and inner primaries on underwings and make rather long tail appear much shorter. Adult females usually have whitish marks below eyes. **Juveniles** are brownish-black and somewhat mottled tawny on uppersides; heavily streaked black on buffy undersides, including dark flank patches; have a distinctive bold face pattern, and white tail with numerous wavy black bands, subterminal band widest. Face pattern includes wide black malar stripes that extend onto sides of neck. Tawny panels on primaries are visible on upper- and underwings of flying birds. Compared to adults, juveniles' wings are narrower and their tails are longer, giving them a different flight silhouette.

Best field marks are flat wings of soaring and gliding birds, long legs of perched birds, white band on upper tail of flying adult and whitish tail with numerous wavy bands of juvenile.

Similar Species:
Zone-tailed Hawk adults (photo ZT04) perched are similar but show whitish lores and all-dark beak and barring on fore-edge of folded wing. Band on uppertail of Zone-tail adult is gray, not white. In flight, their silhouette is much different, with narrower wings and longer tail.
Black Vulture (photos BH01–04) has larger white primary patches, shows whitish legs, soars with wings in a slight dihedral, and lacks white tail band.
Dark-morph buteos show silvery flight feathers that contrast with dark coverts on undersides of wings and have narrower wings.
Range: Breeds in riparian areas of southern Arizona, southwestern New Mexico, and southwestern Texas and usually migrates into Mexico for winter.
Measurements:
      Length: 51–56 cm (54); 20-22 in. (21)
  Wingspan: 102–128 cm (117); 40-50 in. (46)
     Weight: 630–1300 g (950); 1.4-2.9 lb. (2.1)

BH01. **Adult Common Black Hawk.** Note wide wings, white tail band, and small white commas at base of outer primaries. Note retained brownish juvenile underwing coverts. [Mexico, Mar]

BH02. **Adult Common Black Hawk.** Black band on trailing wing edges, white tail band, and orangish cere and base of beak are noticeable. [AZ, July]
BH03. **Juvenile Common Black Hawk.** Note primary panels and whitish tail with wavy black bands. [Mexico, Jan]

BH04. **Juvenile Common Black Hawk.** Overall appears blackish brown with noticeable tawny primary panels and whitish tail with irregular black banding. [AZ, July]

BH05. **Adult female Common Black Hawk.** Overall black, with prominent orangish cere and base of beak and white tail band. Adult female usually has whitish marks below eyes. Wingtips do not reach tail tip. Compare to perched adult Zone-tailed Hawk in ZT04. [AZ, July]

BH06. **Juvenile Common Black Hawk.** Overall blackish-brown, with bold face pattern, especially wide dark malar stripe extending onto neck, and heavily streaked underparts. [AZ, July]

BH07. **Juvenile Common Black Hawk.** Overall blackish-brown. White tail has numerous wavy dark tail bands and wide dark subterminal band. Wingtips do not reach tail tip. Note long legs. [AZ, July]

HARRIS' HAWK *(Parabuteo unicinctus)* Photos HH01–05

Harris' Hawk is a long-legged dark buteonine that has two recognizably different plumages: adult and juvenile. Females are noticeably larger than males. Wingtips of perched birds fall far short of tail tip. **Adults** are dark brown overall except for chestnut leg feathers and wing coverts, white uppertail coverts, and wide white base and terminal band on tail. **Juveniles** are similar to adults, but their underparts are streaked whitish, with breast usually darker than belly; their flight feathers are barred; and their tails have narrow dark barring and have narrower white base and terminal band. Compared to adults, juveniles' wings are a bit narrower and their tails are longer.

Best field marks are chestnut leg feathers and wing coverts, paddle-shaped wings and all-dark underwings of flying birds, wing/tail ratio, and wide white tail tip of adults.

Similar Species:
Dark-morph buteos have silvery flight feathers that contrast with dark coverts on underwings and lack chestnut on legs and wing coverts. Wingtips usually extend to near tail tip on perched buteos.
Red-shouldered Hawk juveniles (photos RS13–14) can appear similar to paler juvenile Harris' but lack large chestnut patches on shoulders and leg feathers. In flight crescent-shaped wing panels are diagnostic.
Range: Resident in desert areas of southern and central Arizona; southern New Mexico, and southern Texas.
Measurements:
      Length 49–59 cm (52); 18–23 in. (20)
    Wingspan: 103–119 cm (108); 40–47 in. (43)
      Weight: 568–1203 g (890); 1.3–2.6 lb (2)

HH01. **Adult Harris' Hawk.** Note paddle-shaped wings and all dark underwings; chestnut coverts are visible in good light. White undertail coverts and base and wide tip of tail are distinctive. [AZ, Nov]

HH02. **Juvenile Harris' Hawk.** Note barred tail and flight feathers, chestnut underwing coverts, and white undertail coverts. Whitish undersides of primaries are paler than rest of underwing and appear as pale panels. [TX, Nov]

HH03. Adult Harris' Hawk. Chestnut upperwing coverts and wide white bands at base and tip of tail are distinctive. White mark on middle of upper tail is tip of growing tail feather. [TX, Oct]

HH04. Adult Harris' Hawk. Dark brown except for chestnut shoulder patches and leg feathers and white tail coverts and tail tip. Note orangish-yellow face skin and cere. [Mexico, Jan]

HH05. Juvenile Harris' Hawk. Is similar to adult, but with streaked underparts, barred (or uniformly chestnut) leg feathers, barred underside of long tail, and yellowish face skin and cere. Note long legs. Wingtips reach only half way down long tail. [TX, Jan]

GRAY HAWK (*Buteo nitidus*)   Photos GH01–06

The Gray Hawk is a small, accipiter-like, long-tailed buteo and has two recognizably different plumages: adult and juvenile. Females are larger than males. Wingtips of perched birds fall far short of tail tip. **Adults** are overall gray with whitish barring on underparts and whitish underwings. Black tail has two white bands. **Juveniles** appear different from adults; they are overall dark brown, with streaked underparts, bold face pattern, barred leg feathers, and long brownish tail with five or more progressively wider, chevron-shaped, dark brown bands. Face pattern consists of white superciliary, dark eye-line, white cheek, dark malar, and white throat.

Best field marks are white U above base of tail, rounded wingtips, pale underwings lacking bold dark border, and wing/tail ratio.

Similar Species:
Broad-winged Hawk juveniles (photos BW04–06,09–11) are similar but have more pointed wingtips, have streaked, not barred, leg feathers, and dark band on trailing edge of wings and lack strong face pattern, especially dark eye-line, and white U above tail. Adult Broad-wings have rufous breasts.
Hook-billed Kite adult males (photos HB01,04) are also gray overall with white barring on underside but have white eyes and paddle-shaped wings that appear dark on undersides.
Range: Breeds in riparian areas of southern Arizona; this population migrates south for winter. A few pairs are resident along lower Rio Grande valley of Texas.
Measurements:
      Length: 36–46 cm (42); 14-18 in. (17)
    Wingspan: 82–98 cm (87); 32-38 in. (34)
      Weight: 378–660 g (524); 13-23 oz (18)

GH01. **Adult Gray Hawk.** Pale underwing, gray body, and long tail with two white bands are distinctive. Wingtips are rather rounded. [Mexico, Jan]

GH02. **Juvenile Gray Hawk.** Rounded wingtips with outer primaries barred, pale underwing lacking dark trailing edge, and long tail with progressively-wider bands near tip are distinctive. [Mexico, Oct]

GH03. **Adult Gray Hawk.** Overall gray coloration and white U above tail are distinctive. [Mexico, Jan]

GH04. **Juvenile Gray Hawk.** Striking face pattern, especially dark eye-line, and white uppertail coverts are diagnostic. [Mexico, Oct]

GH05. **Adult Gray Hawk.** Overall gray coloration, dark eye, and barred underparts are diagnostic. Note bright orange-yellow legs and cere. [Mexico, Jan]

GH06. **Juvenile Gray Hawk.** Striking face pattern, especially dark eye-line, is diagnostic. Outer bands on long tail are progressively wider. Note barred leg feathers. Compare to juvenile Broad-winged Hawks in photos BW09–11. [Mexico, Jan]

RED-SHOULDERED HAWK (*Buteo lineatus*)
Photos RS01–17

The Red-shouldered Hawk, a long-legged woodland buzzard, has two recognizably different plumages, adult and juvenile, and five races that are similar but have recognizably different plumages. Wingtips fall somewhat short of tail tip on perched birds. Sexes are alike in plumage; females average larger than males. The races are **Eastern, *lineatus*; North Florida, *alleni*; South Florida, *extimus*; Texas, *texanus*;** and **California, *elegans*.** The following description is for nominate **Eastern. Adults** have brown heads, backs, and greater upperwing coverts, all somewhat streaked with tawny. Lesser and median upperwing coverts are rufous and form red shoulder of perched birds. Underparts vary from bright to pale rufous (males average brighter) with white barring, and usually, unique to this race, show dark brown streaking on breast. Flight feathers are checkered black and white, with a crescent-shaped white area across primaries. Black tail has three narrow white bands. **Juveniles** have brown heads, back, and upperwing coverts, latter may show a hint of rufous shoulder. Buffy underparts are variously streaked with dark brown. Flight feathers are brown, with a distinctive crescent-shaped tawny panel on upperside of primaries that also shows on back-lighted underwings. Upperside of tail is dark brown with narrow pale bands, usually with a rufous wash at base, but appears pale with darker bands on underside. **North Florida** birds occur south of Eastern ones. **Adults** usually lack dark streaking on breast. **Juveniles** are more heavily marked on undersides with arrow-shaped markings that appear somewhat like barring. Uppersides of secondaries show pale barring, and undertails appear dark with narrow pale bands. **South Florida** birds are the palest overall. **Adults** are pale gray on head, back, and greater upperwing coverts and paler rufous on underparts. **Juveniles** are lightly streaked on underparts, and show pale barring on uppersides of primaries. **Texas** birds are similar to North Florida birds, but **adults** average brighter on underparts. **Juveniles** are almost identical to North Florida juveniles and are not safely separated. **California** birds are the most distinctive, being geographically isolated. **Adults** average brighter on underparts, with, unique to this race, solid rufous breasts and wider white bands in tail. **Juveniles** are quite different from those of other races, being more adult-like, with white crescent-shaped wing panels, pied flight and tail feathers, rufous shoulders, and barred rufous-buff underparts.

Best field mark is the crescent-shaped pale primary wing panels that are visible on upperwings and backlighted underwings of flying birds.

Similar Species:
Broad-winged Hawk juveniles (photos BW09–11) perched are almost identical but have shorter legs, pale brown uppertail with dark bands, and unmarked secondaries. In flight they show pointed wingtips and lack crescent wing panels.
Broad-winged Hawk adults (photos BW01–02, 07–08) also have somewhat rufous underparts and dark tail with white (one wide and one narrow) bands, but lack pale crescent-shaped wing panels and rufous underwing coverts, and pied upperwings and have at most two white bands in tail.
Goshawk juveniles (photo G07) can appear similar when perched. But their wingtips reach only half way down tail, their tails show even-width dark and pale bands, and in flight, they lack pale wing panels.
Northern Harrier juveniles (photos NH03–04) in flight from below resemble adult Red-shouldereds in wing and tail shape and rufous underparts and underwing coverts but show a dark patch of secondaries and coverts on each underwing and lack pale crescent-shaped primary panels.
Range: The borders between races are not well defined, and racial characters become blurred in the contact zones among the first four races. Eastern birds from the north of range migrate south for winter.
Eastern, race *lineatus*: Eastern North America from Minnesota and southeastern Canada

south to northeastern Texas and South Carolina. **North Florida**, race *alleni:* South Carolina south to central Florida and west to central Louisiana. **South Florida**, race *extimus:* Central to southern Florida. (Both Florida races occur together over a large area, with Eastern migrants there also in winter.) **Texas**, race *texanus:* Western Louisiana and eastern Texas south and west along Gulf Coast to south Texas and northeastern Mexico. **California**, race *elegans:* Western California along Pacific Coast barely into southwestern Oregon, and eastward into the Central Valley.

Measurements:

Length:	38–47 cm (42);	15–19 in. (17)
Wingspan:	94–107 cm (101);	37–42 in. (40)
Weight:	460–930 g (629);	1.1–1.9 lb. (1.4)

RS01. **Eastern adult Red-shouldered Hawk.** Eastern adults show wide dark streaking on breast. Note pied flight feathers, rufous underwing coverts, black tail with narrow white bands, and white crescent wing panels. [NY, Mar]

RS02. **Texas adult Red-shouldered Hawk.** Texas adults average brighter than Eastern adults and usually lack dark breast streaks. (Compare with RS01.) [TX, Jan]

RS03. **Adult Red-shouldered Hawk.** Note bold black and white flight feathers and white crescent-shaped primary panels. [TX, Dec]

RS04. **Juvenile Red-shouldered Hawk.** Note tawny crescent primary panels and dark tail with narrow pale bands. [NJ, Nov]

RS05. **California juvenile Red-shouldered Hawk.** Very adult-like : rufous upperwing coverts, pied uppersides of flight feathers, whitish crescent-shaped wing panels, and black tail, with narrow white bands [CA, Aug]

RS06. **Juvenile Red-shouldered Hawk.** Typical juvenile with streaked underparts, crescent-shaped pale wing panels, and dark tail with narrow pale bands. Note wings pushed forward in soar. [TX, Oct]

RS07. **California juvenile Red-shouldered Hawk.** Very adult-like: rufous underwing coverts, crescent-shaped wing panels, barring on flanks, and dark tail with narrow white bands. [CA, Nov]

RS08. **Eastern adult Red-shouldered Hawk.** Eastern adults have dark brown heads, dark breast streaks, and (usually) four white tail bands. [FL, Jan]

RS09. **Texas adult Red-shouldered Hawk.** Texas adults average brighter rufous on underparts, lack dark breast streaking, and usually have three narrow white bands in tail. Pied flight feathers are typical of all adults. Wingtips fall somewhat short of tail tip. [TX, Jan]

RS10. *California adult Red-shouldered· Hawk.* California adults have bright rufous underparts with solid rufous bibs, wider white tail bands, and paler heads. Wingtips fall somewhat short of tail tip. [CA, Nov]

RS11. South Florida Adult Red-shouldered Hawk. Adults in **south Florida** are paler overall, with grayish heads, but still show red shoulder, pied flight feathers, and black tail with narrow white bands. [FL, Jan]

RS12. **Eastern juvenile Red-shouldered Hawk.** Note long legs, dark tail with narrow pale bands, and pale bands on folded secondaries. Wingtips fall somewhat short of tail tip. Compare to perched juvenile Broad-winged Hawk in photo BW11. [VA, Aug]

RS13. **Texas juvenile Red-shouldered Hawk.** Underparts are heavily marked with arrow-shaped markings. Note long legs, dark tail with narrow pale bands, pale bands on folded secondaries, and rufous wash at base of tail. Wingtips fall somewhat short of tail tip. [TX, Jan]

RS14. **North Florida juvenile Red-shouldered Hawk.** Underparts are heavily marked with arrow-shaped markings. Note long legs and dark tail with narrow pale bands. [FL, Jan]

RS15. **South Florida juvenile Red-shouldered Hawk.** Overall paler, with light streaking on underparts. Note long legs, dark tail with narrow pale bands, and pale bands on folded secondaries. [FL, Jan]

RS16. **California juvenile Red-shouldered Hawk.** California juveniles are more adult-like than other juveniles: barred underparts, red shoulder, and pied flight feathers. Wingtips fall somewhat short of tail tip. [CA, Nov]

RS17. **California juvenile Red-shouldered Hawk.** California juveniles are more adult-like than other juveniles: pied flight feathers and blackish tail with white bands. Note pale bands on secondaries. [CA, Nov]

BROAD-WINGED HAWK   (*Buteo platypterus*)

Photos BW01–11

The Broad-winged Hawk, a small woodland buzzard, has four recognizably different plumages: adult and juvenile of both light and (rare) dark morphs. Sexes are alike in plumage; females average larger, but with overlap. Wingtips fall somewhat short of tail tip on perched birds. **Adults** have brown heads, back, and upperwing coverts. White underparts are barred with reddish-brown, lighter on belly. Some adults have solidly marked breast forming reddish-brown bib. Underwings appear uniformly pale with dark tips on outer primaries and wide dark band on trailing edge. Dark tail has one wide and one or two usually obscured narrow white bands. **Juveniles** have brownish backs and pale underparts, with a variable amount of dark streaking from almost none to heavily marked. Underwings are like those of adults except that band on trailing edges is dusky and narrower. Juveniles can show square pale primary panels on backlighted wings. Juvenile's tail is pale brown with many narrow dark bands, subterminal band wider; occasional juveniles can show accipiter-like tail markings of equal-width dark and pale bands. **Dark-morph adults** have dark brown heads, bodies, and wing and tail coverts. Underwings show silvery flight feathers with dark wingtips and dark band on trailing edges. Tail is like that of light-morph adult. **Juvenile dark morph** is like adult except that body and coverts may show tawny streaking and undersides of flight feathers and tail are like those of light-morph juvenile.

Best field marks are pointed wingtips, underwings that appear unmarked except for dark border, and, on adults, dark tail with one wide white band.

Similar Species:
Red-shouldered Hawk (photos RS01–17) is very similar. Perched adults and juveniles are similar to same age Red-shouldered Hawks. See under that species for differences.
Cooper's Hawks (photos C01–06) can appear similar to juvenile Broad-wings, some of which can show equal-width dark and pale tail bands, but Cooper's have short primary projection, lack dark malar stripes, and have short, rounded wings and heavily barred flight feathers.
Goshawk juveniles (photo G07) can appear similar when perched. But they are much larger and their wingtips reach only half way down tail, their tails show irregular tail bands, and in flight, they show rounded, not pointed wingtips.
Range: Breeds in forests of eastern and northern North America west to Alberta and south to east Texas and northern Florida. Winters mostly south of the U.S., but some birds recorded every winter in s. California, along the Gulf Coast, and in s. Florida. Rare dark morph apparently breeds only in Alberta but occurs during migration rarely throughout the West and more often (but still rarely) on the eastern Great Plains.
Measurements:
> **Length:** 34–42 cm (37); 13–17 in. (15)
> **Wingspan:** 82–92 cm (86); 32–36 in. (34)
> **Weight:** 308–483 g (401); 11–17 oz (14)

BW01. **Adult Broad-winged Hawk.** Note pointed wingtips, relatively unmarked underwings with wide dark band on border. Adults have rufous barred underparts and dark tail with one wide white band. Second narrow white band just visible on spread tail. [TX, Oct]

BW02. **Adult Broad-winged Hawk.** Wingtips become sharply pointed when gliding. Note rather straight trailing edge of wing. [NY, Apr]

BW03. **Adult dark-morph Broad-winged Hawk.** All dark brown body and wing coverts and silvery flight feathers typical of dark buteos. Dark tail with one wide white band is diagnostic. Note pointed wingtips. [TX, Oct]

BW04. **Juvenile Broad-winged Hawk.** Unmarked underwings have dark tips and narrow dusky band on trailing edges. Wingtips are pointed. This bird is moderately streaked with a typical clear area in center of breast and hint of Red-tailed Hawk-like belly band. [CT, Oct]

BW05. **Juvenile Broad-winged Hawk.** Wingtips become sharply pointed when gliding. Note that underparts are as heavily streaked as juvenile Red-shouldered Hawk. Relatively unmarked underwings have narrow dusky band on trailing edges. [CT, Oct]

BW06. **Juvenile Broad-winged Hawk.** Backlighted wings show square primary panels. Note fine barring on tips of outer primaries. Tail shows narrow dark bands with subterminal band wider. [MN, Sept]

BW07. **Adult Broad-winged Hawk.** Underparts are barred rufous, lighter on belly. Note single wide white band in tail. [PA, Apr]

BW08. **Adult Broad-winged Hawk.** Variation with solidly marked breast forming bib. Compare to adult Swainson's Hawks in photos SH14-16. Note single wide white band in tail. [TX, Oct]

BW09. **Juvenile Broad-winged Hawk.** Shows pale superciliary, dark malar stripe, and streaked underparts. Note short legs. Compare to juvenile Red-shouldered Hawk in photo RS12. [FL, Jan]

BW10. **Juvenile Broad-winged Hawk.** Some juveniles have completely unmarked underparts. [PA, Aug]

BW11. **Juvenile Broad-winged Hawk.** Shows pale superciliary, dark malar stripe, and brown tail with narrow dark brown bands. Note unmarked secondaries. Wingtips fall somewhat short of tail tip. Compare to juvenile Red-shouldered Hawk in photo RS12. [NJ, Oct]

SHORT-TAILED HAWK   (*Buteo brachyurus*)

Photos ST01-05

The Short-tailed Hawk, a small aerial buzzard that occurs in Florida, has four recognizably different plumages: adult and juvenile of both light and dark morphs. Sexes are alike in plumage; females average larger. All birds show white spot above and beside cere. Light-morph birds share the two-toned underwing pattern of Swainson's Hawks. Wingtips reach tail tip on perched birds, but they seldom perch in the open, usually perching concealed inside tree canopy; most sightings are of flying birds. **Light-morph adults** have dark heads with dark cheeks and white throats. Body and underwing coverts are creamy to white and unmarked, except for small indistinct rufous areas on both sides of upper breast. Flight feathers below are pale grayish, with wide dark terminal band, and contrast somewhat with white coverts and show a whitish oval area in primaries. Undertail is whitish with a wide dark subterminal band and several other narrow, often incomplete, dark bands. **Light-morph juveniles** are similar to adults but have creamier underparts and underwing coverts and usually show some fine dark streaking on sides of upper breast and pale streaking in cheeks. Underwings show two-tone contrast as do those of adults, but secondaries are darker than primaries. Dark band on trailing edge of underwings is narrower than that of adults. Undertail is whitish with seven or more narrow dark bands, subterminal is widest, but much narrower than that of adults. **Dark-morph adults** have heads, bodies, and wing coverts all dark brown. Undersides of their flight feathers are grayish, with wide dark terminal band and whitish oval areas in primaries as in light morph. Undertail is same as that of light-morph adult. **Dark-morph juveniles** are similar to adults but have whitish spotting on belly and underwing coverts. Undersides of flight feathers and tails same as light-morph juvenile.

Their aerial hunting behavior is distinctive, gliding slowly on flat wings with the wingtips curled upward looking intently at the ground below, seldom flapping. When prey is sighted, they fold into a teardrop shape and plummet to the prey. They will kite in a strong wind but do not hover.

Best field marks are pointed wingtips on very long wings and grayish flight feathers with a large whitish oval spot at bases of outer primaries on each underwing.

Similar Species:
Broad-winged Hawks (photos BW01–11) in all plumages have silvery, not grayish, flight feathers. Wingtips do not reach tail tip on perched Broad-wings, which do perch in the open.
Swainson's Hawk (photos SH01–03) light-morph birds have a similar two-toned underwing pattern, but grayish flight feathers are all the same color; Short-tails show whitish ovals at bases of outer primaries.
Range: Peninsular Florida during the summer; northern birds move to southern third of Florida during winter. Dark-morph birds are more common than light-morph ones. Sight record of light-morph vagrant in Rio Grande Valley of Texas.
Measurements:
      Length: 39–44 cm (41); 15–17 in. (16)
   Wingspan: 83–103 cm (93); 32–41 in. (37)
      Weight: 342–560 g (426); 12–20 oz (15)

ST01. **Adult light-morph Short-tailed Hawk.** Pointed wingtips, dark cheeks, and unmarked white under-parts are distinctive. Underwings are two-toned; white coverts contrast with grayish flight feathers, which show whitish oval at base of outer primaries. Adults have whitish tail with a wide dark terminal band and several narrow incomplete dark bands. [FL, Aug]

ST02. **Juvenile light-morph Short-tailed Hawk.** Wingtips are pointed. Two-toned underwing like adult, but darker secondaries contrast also with paler primaries. Undersides are creamy to pale buffy. Some juveniles show fine dark streaking on sides and belly. Whitish undertail has many narrow dark bands; subterminal band usually is a bit wider. Dark cheek usually noticeable even though streaked. [Mexico, Oct]

ST03. Adult dark-morph Short-tailed Hawk. Flight feathers are grayish and show dark terminal band and whitish ovals at base of outer primaries. Whitish tail has wide dark subterminal band. [FL, Jan]

ST04. Juvenile dark-morph Short-tailed Hawk. Dark underwing coverts and belly are spotted with white. Note large whitish ovals at bases of outer primaries, and pointed wingtips. Pale tail has many narrow dark bands, subterminal band is a bit wider. [Mexico, Oct]

ST05. Juvenile dark-morph Short-tailed Hawk. Wingtips become even more pointed when gliding. Paler variant, with more noticeable dark bib and whitish markings on underwing coverts and belly being more streaks than spots. [FL, Jan]

SWAINSON'S HAWK (*Buteo swainsoni*)

Photos SH01–24

The Swainson's Hawk, a large western open country buzzard, is polymorphic; base plumage color of adults varies from whitish to rufous to dark brown to black. Three age classes are recognizable: adult, Basic I (one year old), and juvenile. Sexes are similar in plumage; females tend to be darker and more heavily marked. Females average larger than males. Wingtips reach or slightly exceed tail tip on perched adults, reach or almost reach on juveniles. Adults can show every intermediate plumage from the palest light morph to the darkest dark morph. **Light-morph adults** have dark heads with a large white throat patch and small white patch on forehead and outer lores. Back and upperwing coverts are uniformly dark brown. Males often have grayish faces. Breast is typically rufous or dark brown, forming a dark bib, and paler belly is lightly to heavily barred, sometimes unbarred. Undersides of flight feathers are medium gray, have a wide darker gray terminal band, and contrast with white coverts. Axillaries are marked similarly to belly; underwing coverts are marked less heavily than belly. Pale uppertail coverts form a pale U at base of tail. Tail is medium gray with numerous narrow dark gray bands and a wide dark subterminal band. **Rufous-morph adults** have dark brown breast and paler rufous belly or same shade rufous breast and belly. Underwing coverts vary from whitish to rufous. Other characters are as in light-morph adults. **Dark-morph adults** are dark brown to black overall, except for pale undertail coverts. However, underwing coverts vary from white with a variable amount of rufous mottling to black with a variable amount of whitish mottling. **Light-morph juveniles** have buffy heads with pale superciliary lines and cheeks and some dark streaking and dark back and upperwing coverts with wide tawny-buffy feather edges and mottling. Creamy underparts are variably streaked from light to heavy, usually with a dark patch on each side of upper breast. Underwing is two-toned: creamy coverts contrast with grayish flight feathers, but contrast is less than contrast on underwings of adults. Pale brown tail has numerous narrow equal-width dark bands. Juveniles have narrower wings and longer tails compared to subsequent plumages. **Dark- and rufous-morph juveniles** are similar to light-morph juveniles but their underparts and underwing coverts are a darker buff and are more heavily streaked. One-year-old birds returning in spring are still in juvenile plumage but are usually pale headed from feather wear and fade. During summer, they molt into **Basic I** plumage, which is similar to juvenile plumage but shows wider subterminal band on tail and terminal band trailing edge of underwing. Similarly, two-year-old birds returning in spring are still in Basic I plumage, often with paler heads; adult plumage is acquired by molt during the summer.

Best field marks are long pointed wings and two-toned underwing: grayish flight feathers contrast with paler coverts.

Similar Species:
Red-tailed Hawk juveniles in flight (photos RT17–26) are similar but always show dark patagial marks on underwings, pale wing panels on backlighted underwings, and two-toned upperwings and lack pointed wingtips.
Red-tailed Hawk juveniles perched (photos RT38–40) are similar to juvenile Swainson's with pale scapulars, pale head, and same tail pattern, but their wingtips fall short of tail tip and they have dark belly bands.
Dark-morph buteos of other species have silvery undersides of flight feathers and dark undertail coverts.
Prairie Falcons perched (photos PF06–07) can appear similar to some pale Swainson's but have dark eyes, pale areas behind eyes, and wingtips that reach only half way down tail.

Range: Breeds on prairies and open areas of western North America from southern

Canada to south Texas and central California. Also sparsely in Alaska and northwestern Canada. Most of population migrates to South America, but a few remain in s. California, s. Texas, and s. Florida.

Measurements:

> **Length:**   43–55 cm (49); 17–22 in. (19)
> **Wingspan:** 120–137 cm (128); 47–54 in. (51)
> **Weight:**   595–1240 g (849); 1.3–2.7 lb. (1.9)

SH01. **Adult Swainson's Hawk.** Typical adult with dark bib, unmarked white belly, and two-toned underwing. Note long pointed wings and white throat patch. [CO, Sept]

SH02. **Adult Swainson's Hawk.** Typical adult with dark bib, barred belly, and two-toned underwing. Note long pointed wings and white throat patch. [CO, Sept]

SH03. **Adult Swainson's Hawk.** Adult with heavily barred belly and axillaries. Underwing coverts are lightly marked. [ID, May]

SH04. **Adult rufous-morph Swainson's Hawk.** Typical rufous bird with dark bib and rufous belly. Underwing coverts have a variable amount of rufous. Note white undertail coverts. [CO, Sept]

SH05. **Adult dark intermediate Swainson's Hawk.** Intermediate adult between dark and light, with heavily barred belly and underwing coverts. At a distance appears to be dark. [CO, Sept]

SH06. **Adult dark-morph Swainson's Hawk.** Typical dark adult with all dark body and rufous wing coverts. Note whitish undertail coverts. [CO, Sept]

SH07. **Adult dark-morph Swainson's Hawk.** Dark extreme with all dark body and underwing coverts. Pale undertail coverts are barred. Wingtips can appear more rounded when soaring. [CO, Apr]

SH08. **Basic I Swainson's Hawk.** Basic I plumage similar to juvenile, but new flight and tail feathers have wide subterminal bands. Underwings are two-toned. Note that new secondaries are longer than old juvenile ones. [CO, Sept]

SH09. **Basic I rufous-morph Swainson's Hawk.** Basic I plumage similar to juvenile, but new flight and tail feathers have wide subterminal bands. Like bird in SH08 but more heavily streaked. Note that new secondaries are longer than old juvenile ones. [CO, Sept]

SH10. **Juvenile Swainson's Hawk.** Pale bird with light streaking. Note dark patches on sides of upper breast, narrow subterminal band on tail, narrow terminal band on trailing edge of wings, and two-toned underwing. [CO, Sept]

SH11. **Juvenile Swainson's Hawk.** More heavily marked bird. Note narrow subterminal band on tail and narrow terminal band on trailing edge of wings. [CO, Sept]

SH12. **Juvenile Swainson's Hawk with belly band.** Unusual variation with dark belly band, but told by dark underwings, long pointed wing tips, and white undertail coverts. [CO, Sept]

SH13. Juvenile dark-morph Swainson's Hawk. Dark juveniles have heavily streaked underparts and darkly marked coverts. Note whitish undertail coverts. Dark-morph birds do not show two-toned underwings, but note wing shape. [CO, Sept]

SH14. Adult Swainson's Hawk. Typical adult with rufous breast and unmarked white belly. Note whitish throat patch. Gray face typical of adult males. Wingtips barely exceed tail tip. [CO, Sept]

SH15. **Adult Swainson's Hawk.** Typical adult with dark brown breast and marked belly. Note whitish throat patch. Note wingtips barely exceed tail tip. [OK, June]

SH16. **Adult Swainson's Hawk.** Typical adult with dark brown breast and heavily barred belly. [CO, Sept]

SH17. **Adult rufous-morph Swainson's Hawk.** Typical rufous adult with dark breast and rufous belly. Only rufous buteo with a white throat. [CO, Sept]

SH18. **Adult rufous-morph Swainson's Hawk.** Paler adult lacking a dark bib. Only rufous buteo with a white throat. Note also white undertail coverts. Wingtips extend beyond tail tip. [CO, Sept]

SH19. **Adult dark-morph Swainson's Hawk.** Adult with dark head and body. Note whitish undertail coverts. Wingtips reach tail tip. [CO, Sept]

SH20. **Basic I (two year old) dark-morph Swainson's Hawk.** Basic I bird just returned and before molting into adult plumage. Heavily streaked and not yet uniformly dark. Undertail shows wide dark subterminal band. [ID, May]

SH21. **Juvenile (one year old) Swainson's Hawk.** Juveniles just returned appear quite pale headed before molting into Basic I. Dark subterminal band not wide. Note pale scapular patches. Wingtips reach tail tip. [WY, June]

SH22. **Juvenile (one year old) rufous- or dark-morph Swainson's Hawk.** Juvenile returning from winter grounds appears quite pale headed before beginning molt into Basic I. Juvenile's dark subterminal band same width or barely wider than other dark tail bands. Underparts and leg feathers are more heavily marked than in light morph. [ID, May]

SH23. **Juvenile Swainson's Hawk.** Most juveniles show dark patches on each side of upper breast. Note that wingtips almost reach tail tip [CO, Sept]

SH24. **Juvenile Swainson's Hawk.** Juvenile recently fledged shows pale feather edges, including pale patches on scapulars. Compare to juvenile Red-tailed Hawk in photo RT40. Wingtips almost reach tail tip. [CO, Sept]

WHITE-TAILED HAWK *(Buteo albicaudatus)*

Photos WT01–15

The White-tailed Hawk, a large, long-legged buteo resident of Texas coastal prairies, has three recognizably different plumages: adult, Basic I (subadult), and juvenile. Sexes are alike in plumage; females average larger than males. All soar and glide with wings in a strong dihedral. Wingtips exceed tail tip on perched juveniles and far exceed tip on perched adults and subadults. **Adults** have gray heads with white throats, but throat is dark on young adults. Back and upperwing coverts are gray, except for chestnut lesser coverts, which form shoulder patch on perched birds. Older adults are paler gray; younger ones are blackish-gray. Underparts are white, either unmarked or with fine black or rufous barring on belly, flanks, and leg feathers. Underwings are two-toned: coverts are white and contrast with grayish flight feathers, but secondaries are paler gray than primaries. White tail has wide black subterminal band and five or more faint narrow black bands. **Juveniles** are mostly blackish-brown, with two or three whitish patches on sides of face and a variably sized white breast patch. Underparts vary from overall dark with narrow vertical white slit on breast to mostly white with a Red-tailed Hawk-like dark belly band. Likewise, underwing coverts vary from mostly blackish to whitish. Tail is pale gray with many indistinct fine dark bands. Whitish uppertail coverts form a pale U above base of tail on all but darkest juveniles. Juveniles have noticeably narrower wings and longer tails than do adults and subadults. **Basic I** birds (subadults) are similar in proportions to adults but in plumage to juveniles; however, they lack white face spots, have a noticeable narrow dusky subterminal tail band and wide dusky band on trailing edge of wings, and an adult-like chestnut shoulder patch. Subadults, like juveniles, vary in coloration of underparts from pale to dark.

Best field marks are long pointed wings of flying birds that are shared only by Swainson's Hawks and wingtips far exceeding tail tip on perched birds. Adult's white tail with wide black subterminal band is distinctive.

Similar Species:
Swainson's Hawk (photos SH01–24) light-morph birds also have two-toned underwings, but all of their flight feathers are the same color and they show uniformly dark or streaked breasts. Dark-morph adults are similar to darker juvenile White-tails but are smaller, have noticeable tail banding, including a wide dark subterminal band, and lack white breast patch and white uppertail coverts.
Ferruginous Hawk (photos FH01–02, 05–06, 08–10, 13–15) light-morph birds also have all white underparts like adult White-tails but have all white underwings, white cheeks, and lack wide black subterminal tail band.
Dark-morph buteos of other species are similar to dark juveniles but have silvery flight feathers and lack white U above tail base and wingtips that extend far beyond tail tip.
Range: Resident on Texas coastal prairie from just west of Houston and Galveston to Brownsville. Juveniles wander somewhat outside range.
Measurements:
        Length: 46–58 cm (50); 18–22 in. (20)
  Wingspan: 126–135 cm (129); 49–53 in. (51)
      Weight: 880–1235 g (1022); 1.9–2.7 lb. (2.3)

WT01. **Adult White-tailed Hawk.** Note dark cheeks, white tail with wide black subterminal band, long pointed wings, and two-toned underwings. Secondaries are paler than primaries. Some adults lack barring on underparts. [TX, Feb]

WT02. **Adult White-tailed Hawk.** White on upper-tail coverts extends onto lower back. [TX, Jan]

WT03. **Basic I (Subadult) White-tailed Hawk.** Adult-like but blacker, and tail grayish with narrow subterminal dusky band. [TX, Jan]

WT04. **Basic I (Subadult) White-tailed Hawk.** Similar to juvenile but with all dark head and gray tail with narrow dusky subterminal band. Trailing edges of wings show wide dusky terminal band and pinch in at body. [TX, Jan]

WT05. **Basic I (Subadult) White-tailed Hawk.** Less heavily marked bird than WT04, with more rufous tone to belly and covert markings. Grayish tail has narrow dusky subterminal band. Trailing edges of wings show wide dusky terminal band and pinch in at body. [TX, Dec]

WT06. **Juvenile White-tailed Hawk.** Typical darker juvenile with blackish body and coverts and light grayish undersides of tail and flight feathers. Note small white breast patch, long narrow wings, and pointed wingtips. Trailing edges of wings lack wide dusky terminal band and pinch in at body. [TX, Dec]

WT07. **Juvenile White-tailed Hawk.** Typical paler juvenile with larger white breast patch, more whitish in belly and wing coverts, and almost white undertail coverts. Trailing edges of wings lack wide dusky terminal band and pinch in at body. [TX, Dec]

WT08. **Juvenile White-tailed Hawk.** Even paler juvenile (rare) with juvenile Red-tailed Hawk characteristics of pale underwings and dark belly band on whitish underparts, but it lacks dark patagial marks. Compare to Red-tailed Hawk in photos R17–18. [TX, Jan]

WT09. **Juvenile White-tailed Hawk.** Dark above except for pale U on uppertail coverts. Note white face patch and narrow banding on tail, which is usually not discernible. [TX, Jan]

WT10. **Adult White-tailed Hawk.** Note gray head and upperparts, with chestnut shoulder patches and whitish underparts with fine barring on belly, flanks, and leg feathers. Wingtips far exceed tail tip. White tail has wide black subterminal band. [TX, Jan]

WT11. **Younger Adult White-tailed Hawk.** Young adults are blackish-gray on head and upperparts and have dark throats and heavy barring on underparts. Note long legs. Wingtips far exceed tail tip. White tail has wide black subterminal band. [TX, Jan]

WT12. **Basic I (Subadult) White-tailed Hawk.** Subadults have darker gray upperparts with chestnut patches on shoulder, whitish underparts with dark belly markings, and grayish tails with narrow dusky subterminal band. Wingtips far exceed tail tip. [TX, Dec]

WT13. **Juvenile White-tailed Hawk.** Darker juvenile with dark underparts and small vertical white breast patch. Note long legs. Dark head shows whitish patches on side of face. Wingtips exceed tail tip. Tail has faint narrow dark bands. [TX, Jan]

WT14. **Juvenile White-tailed Hawk.** Paler juvenile with larger white breast patch, white markings on belly, and whitish leg feathers. Dark head shows whitish patches on side of face. Wingtips exceed tail tip. Tail has faint narrow dark bands. [TX, Jan]

WT15. **Juvenile White-tailed Hawk.** Dark head shows whitish patches on side of face. Dark upperparts show some faint rufous markings. Wingtips exceed tail tip. Tail has faint narrow dark bands. Note lack of chestnut shoulder patch of subadults and adults. [TX, Jan]

ZONE-TAILED HAWK   (*Buteo albonotus*) Photos ZT01–05

The Zone-tailed Hawk, a large blackish buteo breeding in hills and mountains of the Southwest, has two recognizably different plumages: adult and juvenile. Sexes are almost alike in plumage; females average larger than males. Wingtips reach or slightly exceed tail tip on perched birds. **Adults** are blackish-gray overall, except for silvery undersides of flight feathers and tail bands, which are white on underside and gray on upperside. Undersides of flight feathers are the most heavily barred of the buteonines. Adult males have one wide and one narrow white tail band; adult females have one wide and two narrow white bands. **Juveniles** are similar to adults but overall color is more brownish black and underside of tail is silvery with many narrow dark bands, subterminal band wider; upperside of tail appears uniformly dark, but close in good light shows faint banding. Many juveniles show small white spots on body or wing coverts.

Best field mark is mimicry of Turkey Vulture; birds' overall shape and habit of soaring, gliding, and teetering with wings in a strong dihedral makes them hard to distinguish from Turkey Vultures.

Similar Species:
Turkey Vultures (photos TV01–07) appear almost identical in flight but have red heads (adults) and lack barring and dark subterminal band on underwings, tail bands, and yellow cere.
Common Black Hawk adults (photos BH01–02, 05) are almost identical when perched but show white tail band on upperside of tail, orangish patch on lores and base of beak, have longer legs, and lack barring on fore-edge of folded primaries. In flight their broad wings give them a quite different shape from the Turkey Vulture-like appearance of Zone-tails.
Dark-morph buteos of other species, especially adult Harlan's Red-tail (photos RT10–12) and adult male Rough-legged Hawk (photo RL06) can appear similar. Harlan's have a broader wing, usually have whitish tail and whitish streaking on breast, and show reduced barring on bases of outer primaries. Adult male dark Rough-legs are similar in wingshape but have many narrow white bands in tail and reduced barring on bases of outer primaries. Both overlap in range with Zone-tails only in winter, when most Zone-tails have gone south.
White-tailed Hawk juveniles (photos WT06, WH13–15) can appear similar to juvenile Zone-tails but their tails have less noticeable banding and lack wider subterminal band and they usually show a white breast patch.
Red-tailed Hawk dark-morph juveniles (photos RT24, 44) can appear similar to juvenile Zone-tails but are browner overall, have a wider wings that show pale panels on outer half, and lack white spots and barring on bases of inner primaries. When perched, their wingtips fall short of tail tip.
Range: Zone-tailed Hawks breed in hills and mountains over most of Arizona and New Mexico and in southwestern Texas. Most leave for winter, but winter records exist, even outside range in s. California and s. Texas.
Measurements:
Length: 48–56 cm (51); 19–21 in. (20)
Wingspan: 121–140 cm (129); 48–55 in. (51)
Weight: 610–1080 g (830); 1.3–2.4 lb. (1.8)

ZT01. **Adult male Zone-tailed Hawk.** Overall blackish-gray with heavily barred flight feathers, yellow cere, and white tail bands. Male has one wide and one narrow band. Silhouette and flight mimic those of Turkey Vulture. [NM, July]

ZT02. **Adult female Zone-tailed Hawk.** Overall blackish-gray with heavily barred flight feathers, yellow cere, and white tail bands. Female has one wide and two narrow bands. Silhouette and flight mimic those of Turkey Vulture. [NM, July]

ZT03. **Juvenile Zone-tailed Hawk.** Overall black with heavily barred flight feathers, yellow cere, and silvery tail with numerous narrow dark bands; subterminal band wider. May or may not show white spots on body and underwing coverts. [AZ, Aug]

ZT04. **Adult Zone-tailed Hawk.** Overall blackish-gray. Note yellow cere and white tail band. Beak is all dark and wingtips extend past tail tip. Compare with perched Common Black Hawk in photo BH05. [AZ, July]

ZT05. **Juvenile Zone-tailed Hawk.** Overall black with a few small white spots. Note yellow cere and silvery undertail with dark bands. Beak is all dark and wingtips extend almost to tail tip. Note wide dark subterminal band. [AZ, Aug]

RED-TAILED HAWK   (*Buteo jamaicensis*)   Photos RT01–46

The Red-tailed Hawk, a large, widespread and common buteo, has many recognizably different plumages, owing to a variety of color morphs and age and racial variation; however, major differences are among color morphs; racial variations among adult and juvenile plumages are relatively minor, with the exception of the Harlan's Hawk. Sexes are alike in plumage; females average larger than males. Wingtips reach tail tip on perched adults, except for Eastern and Harlan's, and fall short of tail tip on perched juveniles. Red-tails soar with wings in a medium dihedral and glide with wings nearly level. Juveniles have noticeably narrower wings and longer tails than do adults. **Juveniles of all races and morphs** have two-toned upperwings; paler outer half (primaries and primary coverts) contrasts with darker inner half. They all also have light brown tails with many equal-width narrow dark brown bands; however, outermost is often a bit wider than others. **Typical light-morph adults** have dark brown head, back, and upperwing coverts. Scapulars are usually heavily spotted buffy, forming a noticeable V on back of perched birds. Pale underparts often have dark markings across belly forming dark belly band and rufous wash on sides of upper breast and neck. Underwings are pale with obvious dark patagial marks. Buffy uppertail coverts form a pale U above base of tail. Rufous tail usually shows a narrow dark brown subterminal band. Eye is dark but often pale on young adults. **Typical light-morph juveniles** have brown head, back, and upperwing coverts. Scapulars are usually heavily spotted white, forming a noticeable V on back of perched birds. Whitish underparts have dark markings across belly forming dark belly band, usually wider and darker than those of corresponding adult. Underwings are pale with obvious dark patagial marks. Pale uppertail coverts form a pale U above base of tail. Eye is pale brown.

Races are **Eastern, Western, Fuertes, Florida,** and **Harlan's**. Typical birds of each race are described below. There is much intergrading of plumages among races and color morphs.

Eastern. Typical **adults** have medium brown head and upperparts and white throats, show less rufous on underparts, and occasionally lack belly bands. Wingtips fall short of tail tip. Typical **juveniles** have fairly pale heads, medium brown upperparts, and medium width dark belly bands. Wingtips fall a ways short of tail tip. Northeastern Canadian birds differ; they are more similar to Western birds.

Western. Dark and rufous morphs occur only in Western birds and only as a small fraction. Typical **light-morph adults** have darker brown head and upperparts, dark throats, a rufous wash on underparts, and wider dark belly bands, which often include dark barring, a feature rarely found on other races. Rufous tail may show additional narrow dark bands. Typical **light-morph juveniles** have dark brown head and upperparts and wide dark brown belly band. Wingtips reach almost to tail tip. Typical **dark-morph adults** have dark brown to jet black (rare) body and coverts with typical rufous tail, which usually has a wider subterminal band and many narrower bands. Typical **rufous-morph adults** are similar to brownish dark-morph adults but with a large rufous patch on breast and rufous legs and underwing coverts; dark patagial marks are often noticeable. They may or may not show pale markings on scapulars. Typical **dark-morph juveniles** are overall dark brown with a variable amount of rufous-buff streaking on body and underwing coverts, from none to heavy. Typical **rufous-morph juveniles** are similar to Western light-morph juveniles but are more heavily marked, with wider belly band and heavy dark streaking on breast. Dark patagial mark on underwing usually obscured by dark markings. All dark- and rufous-morph juveniles show the characteristic two-toned upperwings.

Fuertes. Typical **adults** lack dark belly bands but usually show faint barring on flanks. They have dark (Southwest) or white (Texas and Oklahoma) throats. Typical **juveniles** are almost identical to Eastern juveniles, except they have narrower, less heavily marked belly bands and their wingtips almost reach tail tip on perched birds.

Florida. Smallest race. Typical **adults** are almost identical to Western light-morph adults. Typical **juveniles** are almost identical to Eastern light-morph juveniles.

Krider's. Krider's is not a race but a whitish color morph of Eastern Red-tail that apparently breeds in the northern Great Plains. Typical **adults** have upperparts heavily mottled with white, often including an all-white head. Underparts are completely white, with little or no rufous wash on neck and never show a belly band or flank barring. White underwings usually show faint patagial marks. Tail is usually basally white and outer half pale rufous. **Typical juveniles** have upperparts heavily mottled with white, often including an all-white head. White underparts usually show a faint belly band. Uppersides of primaries and primary coverts are much whiter than inner wing, forming strongly two-toned upperwings. White underwings usually show faint patagial marks and narrow dark barring on tips of outer primaries. Dark bands in tail are narrower than those of other races; basal half of tail usually much whiter than outer half, which sometimes lacks banding.

Harlan's. Harlan's occurs mainly in a dark morph; only a small fraction of birds are light-morph. Typical **dark-morph adults** are coal-black overall. They usually show whitish streaks on breast but seldom show white markings on scapulars. Whitish to grayish tail shows either dark mottling, usually with a smudgy darker terminal band, or dark banding and sometimes has a rufous cast because of intergrading with other Red-tails. Wingtips fall short of tail tip. Typical **dark-morph juveniles** are similar to Western dark-morph juveniles but are overall blackish-brown, usually show whitish breast streaking and whitish markings on scapulars and have wide wavy dark tail bands and outermost primaries barred to tips. Wingtips fall somewhat short of tail tip. Typical **light-morph adults** have blackish-brown upperparts with some white scapular markings and very white underparts that lack the rufous on other races. Dark belly band is light to moderate, and dark patagial marks are pronounced. Tails are like those of dark-morph birds. Typical **light-morph juveniles** have blackish-brown upperparts and whitish underparts with dark belly band. Tail has wavy dark banding and undersides of outer primaries are barred to tips.

Best field marks for light-morph birds are dark patagial marks on underwings of all ages, red tails of adults, and two-toned upperwings of juveniles. Belly bands are not diagnostic, as other species can show this character and not all light-morph Red-tails do so. Best field marks for dark-morph birds are tail pattern and color and two-toned upperwings of juveniles.

Similar Species:
Ferruginous Hawk (photos FH01–14) can appear similar; see under that species for differences.
Rough-legged Hawk (photos RL01–20) can appear similar; see under that species for differences.
Swainson's Hawk (photos SH01–24) can appear similar; see under that species for differences.
Dark-morph buteos of other species are similar but lack rufous tails of adult Red-tails or two-toned upperwing pattern and even-width narrow dark tail bands of juveniles.
Range: The Red-tailed Hawk is one of the most widespread of our raptors, occurring almost everywhere in North America. **Eastern** birds range from the Great Plains east of the Rocky Mountains from central Canada south to the Gulf coast and northern Florida. Northern birds move south for winter. **Florida** birds are resident on the Florida peninsula. **Western** birds occur from the Rocky Mountains westward. **Harlan's** breeds in Alaska and northwestern Canada and winters in southeastern Great Plains and sparingly throughout West. **Fuertes** is resident in Oklahoma and Texas on the Great Plains, as well as in Arizona and New Mexico and southward.
Measurements:
      Length: 45–55 cm (50); 17–22 in. (19)
  Wingspan: 110–141 cm (125); 43–56 in. (49)
     Weight: 710–1550 g (1082); 1.5–3.3 lb. (2.4)

RT01. **Adult Eastern Red-tailed Hawk.** Rufous tail and dark patagial marks. Eastern adults usually have white throats and may or may not show belly band. [TX, Dec]

RT02. **Adult Florida Red-tailed Hawk.** Told by rufous tail and dark patagial marks. Averages smaller in size. Similar to Western adults, with dark throat, rufous wash on underparts, and multiple tail bands. [FL, June]

RT03. **Adult Western Red-tailed Hawk.** Told by rufous tail and dark patagial marks. Darker than Eastern adult with more rufous wash on underparts and coverts, dark throat, and often multiple tail bands. [CO, Nov]

RT04. **Adult Western (Pacific Northwest) Red-tailed Hawk.** Adult Western Red-tails from Pacific Northwest have a heavier rufous wash on underparts and coverts. [CO, Oct]

RT05. **Adult Fuertes Red-tailed Hawk.** Adults from southern Great Plains and Southwest are similar to Eastern adults, but lack a noticeable belly band. They may (Southwest) or may not (s. Great Plains) have a dark throat. [TX, Dec]

RT06. **Adult Krider's Red-tailed Hawk.** Characterized by white head, whitish, relatively unmarked underparts and underwing coverts that lack rufous wash, and pale rufous tail that is often whitish on base. Patagial marks are usually small but noticeable. [TX, Jan]

RT07. **Adult Krider's Red-tailed Hawk.** Note white head, white base of tail, and whitish mottling on back and upperwing coverts. [TX, Jan]

RT08. **Adult rufous-morph Red-tailed Hawk.** Note rufous patch on breast, solid dark brown belly band, rufous and dark brown underwing coverts, and, usually, multiple bands in rufous tail. Dark patagial marks noticeable. [AZ, Feb]

RT09. **Adult dark-morph Red-tailed Hawk.** Underparts and wing coverts are uniformly dark brown to black. Rufous tail usually has multiple tail bands. [CA, Nov]

RT10. **Adult Harlan's Red-tailed Hawk.** Overall color is blackish, usually with whitish streaking on breast, and whitish to grayish, darkly mottled gray tail that usually shows smudgy dusky terminal band. [MO, Dec]

RT11. **Adult Harlan's Red-tailed Hawk.** Variation with darker gray mottled tail with dusky subterminal band and smaller whitish breast patch. Outer primaries have dark barring near tips. [IA, Oct]

RT12. **Adult Harlan's Red-tailed Hawk.** Variation with banded tail. This individual lacks white breast patch. [CO, Nov]

RT13. **Adult Light-morph Harlan's Red-tailed Hawk.** Note whitish underparts and lack of rufous coloration on head and sides of breast. Underwing shows dark patagial mark. Typical Harlan's tail. [LA, Nov]

RT14. **Adult Light-morph Harlan's Red-tailed Hawk.** Appears much like other light-morph Red-tails but with Harlan's tail. Note few pale markings on scapulars. [AZ, Nov]

RT15. **Adult partial albino Red-tailed Hawk.** Typical bird showing mix of white and normal feathers. Eye, leg, and talon color as in normal adults. Partial albinos occur throughout range but only as adults. [AZ, Nov]

RT16. **Adult partial albino Red-tailed Hawk.** This bird is almost completely white, but one dark lesser underwing covert on left wing. Eye, leg, and talon color as in normal adults. Partial albinos occur throughout range but only as adults. [VA, Apr]

RT17. **Juvenile Eastern Red-tailed Hawk.** Note dark patagial mark and pale brown tail with many narrow dark bands. Throat is white in Eastern birds. Belly band not diagnostic. [MN, Sept]

RT18. **Juvenile Western Red-tailed Hawk.** Like Eastern juvenile but overall darker and more heavily marked, with dark throat. [CO, Sept]

RT19. **Juvenile Eastern Red-tailed Hawk.** Best field mark for all juvenile Red-tails is pale outer half of upperwing, including both primaries and primary coverts. Compare upperwings with those in photo FH05 and in photo RL10. [NY, Apr]

RT20. **Juvenile Fuertes Red-tailed Hawk.** Like Eastern juvenile but with reduced belly band. May or may not have dark throat. [TX, May]

RT21. **Juvenile Krider's Red-tailed Hawk.** Note white crown, whitish primaries and primary coverts, heavy whitish mottling on coverts, and whitish base of tail. [TX, Dec]

RT22. **Juvenile Krider's Red-tailed Hawk.** Whitish almost unmarked below, with faint belly band and patagial marks. Outer primaries are barred; compare to those of juvenile Ferruginous Hawk in photo FH06. Note white head. [TX, Dec]

RT23. **Juvenile rufous-morph Red-tailed Hawk.** Not rufous color like adult but underparts with heavy belly band, heavily marked wing coverts, and streaked breast. Note typical juvenile tail with many narrow dark bands. Patagial marks on underwings usually obscured by heavy markings. [AZ, Nov]

RT24. **Juvenile dark-morph Red-tailed Hawk.** Solid dark brown body and wing coverts, but often streaked with rufous-buff. Note typical juvenile tail; bands usually wider than in light and rufous morphs. [CO, Dec]

RT25. **Juvenile Harlan's Red-tailed Hawk.** Overall color blackish. Similar to rufous-morph juvenile but with barring on outer primary tips. [CO, Nov]

RT26. **Juvenile Harlan's Red-tailed Hawk.** Shows typical juvenile Red-tail pattern of pale outer wing, including primary coverts, that contrast with darker inner wing. [CO, Nov]

RT27. **Adult Eastern Red-tailed Hawk.** Eastern adults have white throats and wingtips fall short of tail tips; drooped wings appear longer. Note rufous tail and whitish markings on scapulars. Older adults have dark brown eyes. [TX, Jan]

RT28. **Adult Western Red-tailed Hawk.** Western adults are heavily marked on underparts with dark throats. On Western adults, wingtips reach tail tip. [CA, Nov]

RT29. **Adult Western Red-tailed Hawk.** On Western adults, wingtips reach tail tip. Note multiple narrow dark tail bands and reduced pale scapular markings. Eye color is pale on younger adults. [CO, Sept]

RT30. **Adult Fuertes Red-tailed Hawk.** Like Eastern Red-tail but with little or no belly band. Wingtips do not reach tail tip and throat is usually white on s. Great Plains. [TX, Dec]

RT31. **Adult Fuertes Red-tailed Hawk.** Like Eastern Red-tail but with little or no belly band. Wingtips reach tail tip and throat is usually dark in Southwest. [AZ, Nov]

RT32. **Adult Krider's Red-tailed Hawk.** Note whitish head, heavy whitish markings on back and coverts, and whitish base of pale rufous tail. Wingtips fall short of tail tip, as on normal Eastern Red-tails. [LA, Dec]

RT33. **Adult rufous-morph Red-tailed Hawk.** Dark head and wide dark belly band and rufous breast are typical of rufous-morph adult. [CO, Oct]

RT34. **Adult dark/rufous intergrade Red-tailed Hawk.** Overall dark brown with some rufous on breast, with rufous tail with dark bands. [AZ, Nov]

RT35. **Adult dark-morph Red-tailed Hawk.** Overall dark brown to black, with rufous tail with dark bands. [KS, Oct]

RT36. **Adult Harlan's Red-tailed Hawk.** Overall blackish. Alternate tail pattern: whitish with narrow dark bands, subterminal wider. [CO, Nov]

RT37. **Adult Harlan's Red-tailed Hawk.** Overall blackish with whitish streaking on breast. First adult plumage; tail has both adult and juvenile feathers. [CO, Nov]

RT38. **Juvenile Eastern Red-tailed Hawk.** Back brownish with whitish mottling on scapulars. Wingtips do not reach tail tip. Brown tail has multiple narrow dark brown bands, all the same width. Juveniles have pale eyes. [TX, Dec]

RT39. **Juvenile Western Red-tailed Hawk.** Like Eastern juvenile but darker brown with dark throat and heavier belly band. [CA, Nov]

RT40. **Juvenile Western Red-tailed Hawk.** Wingtips almost reach tail tip on Western juveniles. [CA, Nov]

RT41. **Juvenile Krider's Red-tailed Hawk.** Like normal juvenile Eastern Red-tails but with whitish head and mottling on back and wing coverts. Wingtips fall short of tail tip. [LA, Feb]

RT42. **Juvenile Harlan's Hawk.** This bird is probably a light-morph juvenile. Note pale head, heavily spotted upperparts, and chevron-shaped tail bands. Wingtips fall quite short of tail tip. [MO, Nov]

RT43. **Juvenile rufous-morph Red-tailed Hawk.** Note heavy dark belly band and dark streaking on breast. [AZ, Nov]

RT44. **Juvenile dark-morph Red-tailed Hawk.** All dark brown overall, often with rufous streaking on underparts. [CO, Nov]

RT45. **Juvenile Harlan's Red-tailed Hawk.** Like juvenile rufous-morph but blacker overall, with whitish streaking on underparts [CO, Nov]

RT46. **Juvenile Harlan's Red-tailed Hawk.** Darkish brown back with chevron shaped dark tail bands. May show whitish on scapulars. Wingtips fall somewhat short of tail tip. [CO, Nov]

FERRUGINOUS HAWK (*Buteo regalis*)   Photos FH01–17

The Ferruginous Hawk, our largest buteo, is resident in arid habitats of the West. It has four recognizably different plumages: adult and juvenile of light and dark morphs. Dark-morph birds constitute less than ten per cent of population. Sexes are alike in plumage; females average larger than males. Adults can have either pale or dark eyes; juveniles have pale ones. They soar with wings in a strong dihedral and glide with them in a modified dihedral. All show whitish primary panels on upperwings that do not include primary coverts. Wingtips fall somewhat short of tail tip on perched juveniles but almost reach tip on adults. Undersides of outer primaries show narrow dark tips; rest of flight feathers show narrow dark (dark-morph adults) or dusky tips. **Light-morph adults** have heads that vary from pale gray to brown, usually with noticeable dark eye-lines. Back and upperwing coverts are mostly chestnut. Whitish underparts have little to lots of rufous barring. Underwing coverts are likewise variably marked rufous. Leg feathers are usually rufous or barred rufous and white and form a noticeable V on belly of flying adults. Tail is rufous, white, or gray, or a mixture of these. **Dark-morph adults** have dark brown heads, backs, underparts, and coverts. Breast, belly, wing coverts, and tail coverts can show a mixture of rufous; breast can also show whitish streaking. Tail is entirely gray. **Light-morph juveniles** have heads that vary from pale to dark brown, usually with noticeable dark eye-line. Their backs and upperwing coverts are dark brown. Whitish underparts usually show some dark spotting. Leg feathers are not dark but white and usually show some dark spotting. Brown tail shows faint dark banding and has white on base of upperside. White uppertail coverts have large dark spots and form pale U above base of tail. **Dark-morph juveniles** have dark brown heads, backs, and coverts, but head and breast have a rufous-tawny cast and contrast with darker belly and underwing coverts. Tail appears pale below with dusky terminal band and all dark above, but may show some faint banding when seen in good light.

Best field marks are long tapered wings with narrow dark tips on outer primaries, broad chest, large head with wide gape, feathered legs, and dark legs forming a V on undersides of flying adults and dark eye-line of perched light-morph birds.

Similar Species:
Red-tailed Hawks (photos RT01–46). Light-morph Red-tails always show dark patagial marks on underwings. Adults usually have a dark subterminal band in rufous tail. Harlan's Red-tails are similar to dark-morph Ferrugs but have darker, wider band on trailing edge of wing and dusky subterminal band on tail and lack pale primary panels on upperwings.

Rough-legged Hawk dark-morph juveniles (photos RL12–13) can be almost identical to dark juvenile Ferrugs but have white foreheads and wide black tips on undersides of outer primaries and lack white commas on wrist.

Dark-morph buteos of other species show wider dark tips of outer primaries and have different tail patterns.

Range: Breeds from Great basin east to western Great Plains, north to Canadian prairies, south to northern Arizona and New Mexico. They move southward, westward, and eastward in winter.

Measurements:
> **Length:** 50–66 cm (59); 20–26 in. (23)
> **Wingspan:** 134–152 cm (143); 53–60 in. (56)
> **Weight:** 980–2030 g (1578); 2.2–4.5 lb. (3.5)

FH01. **Adult Ferruginous Hawk.** Note long tapered wings, dark legs forming a V on belly, narrow black wingtips, and rufous unbanded tail. [CO, Feb]

FH02. **Adult Ferruginous Hawk.** More heavily marked adult: underwing coverts are heavily marked chestnut, and belly is heavily barred chestnut. [ID, May]

FH03. **Adult dark-morph Ferruginous Hawk.** Breast is mostly rufous, and belly and wing coverts are brownish-rufous. Tail is uniformly gray, but appears whitish on underside. Note narrow black tips on primaries and white wrist comma on lower wing. [OK, Jan]

FH04. **Adult Ferruginous Hawk.** Note whitish primary panel and dark primary coverts and chestnut upperwing coverts. Upper tail is rufous and white. Note narrow black tips on outer primaries. [CO, Jan]

FH05. **Juvenile Ferruginous Hawk.** Upperparts are dark brown. Primary patch does not include dark primary coverts (compare to photos RT19 and RL09). White uppertail coverts have large black spots. [CO, Jan]

FH06. **Juvenile Ferruginous Hawk.** Overall whitish with few markings. Note dark flank patches, long tapered wings, lack of dark patagial mark, and whitish unbanded base of tail. Outer primaries are not barred. Compare to those of juvenile Krider's Red-tailed Hawk in photo RT22. [CO, Jan]

FH07. **Juvenile dark-morph Ferruginous Hawk.** Overall dark brown but with paler rufous-tawny cast to head and breast. Note long tapered wings, narrow black tips on outer primaries, and white wrist commas. Compare to juvenile dark-morph Rough-legged Hawk in photo RL12. [CO, Oct]

FH08. **Adult Ferruginous Hawk.** Dark rufous legs are feathered to toes. Note chestnut upperwing coverts, dark eye-line, and large gape. [CO, Feb]

FH09. **Adult Ferruginous Hawk.** Lightly marked individual with whitish leg feathers that would not form a dark V on belly when in flight. Note dark eyeline. [TX, Nov]

FH10. **Adult Ferruginous Hawk.** Back and upperwing coverts are mostly chestnut. Color of upper tail is variable, white, rufous, gray, or a mixture of them. [KS, Oct]

FH11. **Adult dark-morph Ferruginous Hawk.** Underparts mixture of rufous and dark brown. Leg feathers are dark. Note dark forehead and compare to dark Rough-legs in photos RL18–19. Note large gape. Undertail is whitish. [CO, Dec]

FH12. **Adult dark-morph Ferruginous Hawk.** Head and back are blackish-brown. Note dark forehead and compare to dark Rough-legs in photos RL18–19. Upper tail is gray. Wingtips fall just short of tail tip. [CO, Feb]

FH13. **Juvenile Ferruginous Hawk.** Year-old juvenile shows worn plumage. Note dark eyeline. Legs are whitish. [WY, June]

FH14. **Juvenile Ferruginous Hawk.** Back is dark brown, usually lacking any chestnut. Note dark eye-line. Wingtips fall somewhat short of tail tip. [CO, Dec]

FH15. **Fledgeling Ferruginous Hawk.** Recently fledged birds usually have a buffy wash on breast that fades by fall. Note that legs are feathered to toes. [CO, July]

FH16. **Juvenile dark-morph Ferruginous Hawk.** Head, body, and wing coverts are dark brown. Note dark forehead and compare to dark juvenile Rough-leg in photos RL19, 21. Wingtips fall somewhat short of tail tip. [CA. Nov]

FH17. **Juvenile dark-morph Ferruginous Hawk.** Overall dark brown but with paler rufous-tawny cast to head and breast. Note dark forehead and compare to dark juvenile Rough-leg in photos RL19, 21. [CO, Feb]

ROUGH-LEGGED HAWK   (*Buteo lagopus*)

Photos RL01–21

The arctic Rough-legged Hawk, a large, long-winged buteo, is a winter visitor to southern Canada and northern U.S. It has six recognizably different plumages: adult male type, adult female type, and juvenile of both light and dark morphs. Dark-morph birds make up to twenty-five to forty per cent of the population in the East but only ten per cent or less in the West. Adults of one sex may have plumage characters of the other sex. Females average larger than males. They soar with wings in a medium dihedral and glide with wings in a modified dihedral; they often hover. Wingtips reach tail tip on perched juveniles and reach or exceed tip on adults. Adults have dark eyes; juveniles have pale ones. **Adults of both morphs** have a wide black terminal band on trailing edges of underwings, wide black subterminal band on undertails, and dark eyes. **Juveniles of both morphs** have pale primary panels on upperwings (but dark primary coverts), dusky bands on undertail and trailing edge of underwing, and pale eyes. **Light-morph adults** usually show a pale U between breast and belly markings and barring on leg feathers. **Light-morph adult males** have grayish back and upperwing coverts, breast usually marked as heavy or heavier than belly, barred flanks and leg feathers, and usually, multiple dark bands in tail. **Light-morph adult females** have brown back and upperwing coverts and belly more heavily marked than breast, non-barred flanks and heavily marked leg feathers, and usually only one black tail band. **Light-morph juveniles** are like adult females, but with reduced markings on breast, underwing coverts, and leg feathers and with narrower dusky band on underwings and undertails. All **dark-morph birds** except darkest adult males have white forehead. Carpal patch is always black and contrasts with dark brown underwing coverts of dark brown birds. **Dark-morph adult males** are usually overall jet black but can be dark brown and have three or four narrow white bands in black or dark brown tail; whitish bands are always visible on uppersides of tails. **Dark-morph adult females** are overall dark brown with one wide black subterminal band on undertail. Some birds have pale heads. Uppertails usually show no pale banding but can show faint grayish bands. **Dark-morph juveniles** are like adult females but with narrower dusky bands on underwings and undertails. Some birds have pale heads. Uppertails usually show no pale banding, but some birds show pale gray bands. Some adult females and juveniles are **intermediate**, with characters of both color morphs.

Best field marks are dark carpal patches on underwings of light-morph birds and brown dark-morph ones, completely feathered legs, and white forehead and tail pattern of dark-morph birds.

Similar Species:
Ferruginous Hawk dark-morph juveniles (photos FH07, 16–17) can appear almost identical to juvenile dark Rough-legs but have white wrist commas on underwings, dark foreheads, rufous-tawny cast to head and breast, and narrow black tips on outer primaries. Light-morph Ferrugs lack dark carpal patches.
Harlan's Red-tailed Hawk adults (photos RT10–14) can appear almost identical to adult dark-morph Rough-legs but have white or gray on uppertail, whitish patch on breast (usually), and lack wide clear-cut black subterminal tail band.
Dark-morph buteos of other species are separated by tail patterns.
Northern Harriers (photos NH01–08) have white uppertail coverts, not on tail base as in light-morph Rough-legs.
Range: Breeds on Arctic tundra and mixed tundra-boreal forest from western Alaska to Newfoundland. Winters in a wide belt from southern Canada to mid-latitudes of the U.S.
Measurements:
        Length: 46–59 cm (53); 18-23 in. (21)
    Wingspan: 122–143 cm (134); 48-56 in. (53)
       Weight: 745–1380 g (1026); 1.6-3.0 lb. (2.2)

RL01. **Adult male Rough-legged Hawk.** Adults have wide dark terminal band on trailing edges of wings and usually show a pale U between breast and belly. Males usually have multiple dark tail bands, breast equally or more heavily marked than belly, flanks and leg feathers that are always barred, and dark carpal patches that are often not solidly black. [CO, Dec]

RL02. **Adult male Rough-legged Hawk.** Some adult males lack markings on belly but usually show a little barring on flanks and legs. Dark breast markings form a bib. [CO, Dec]

RL03. Adult female Rough-legged Hawk. Adults have wide dark terminal band on trailing edges of wings and usually show a pale U between breast and belly. Females usually have a single wide dark subterminal tail band, dark carpal patches, and belly more heavily marked than breast. Belly may show a solid dark band. [TX, Jan]

RL04. Adult male Rough-legged Hawk. Adults lack pale primary patches on upperwings. Note multiple dark tail bands and grayish cast to back and upperwing coverts. [CO, Mar]

RL05. Adult female Rough-legged Hawk. Adults lack pale primary patches on upperwings. Note dark subterminal band on uppertail and brownish back. [CO, Jan]

RL06. **Adult male dark-morph Rough-legged Hawk.** Overall jet black, but some adult males are dark brown. Black tail has three or four narrow white bands. Underwings of adults have wide dark terminal band. [CO, Dec]

RL07. **Adult female dark-morph Rough-legged Hawk.** Overall dark brownish. Undertail is whitish with wide dark subterminal band. Undersides of silvery flight feathers of adults have wide dark terminal band on trailing edge of wings. [CO, Dec]

RL08. **Adult female dark-morph Rough-legged Hawk.** Upperside of tails of dark-morph birds lack white bases. Adults lack pale primary patches on upperwings. [CO,Dec]

RL09. **Juvenile Rough-legged Hawk.** Juveniles show whitish patches on uppersides of primaries; primary coverts are dark. Uppertail is white on base with dusky band on tip. Uppertail coverts are also white. [CO, Dec]

RL10. **Juvenile Rough-legged Hawk.** Dark carpal patches. Juveniles have narrow dusky subterminal band on trailing edge of wings, dusky subterminal band on undertail, and solid belly band. [CO, Jan]

RL11. **Intermediate juvenile Rough-legged Hawk.** A few juveniles have characters of both light and dark morphs. Head, legs, and underwing coverts are light on this bird, but breast and axillaries are dark. [MN, Oct]

RL12. **Juvenile dark-morph Rough-legged Hawk.** Like dark adult females except for narrow dusky band on trailing edge of underwings and dusky band on tip of tail. Note dark head. [CO, Apr]

RL13. **Juvenile dark-morph Rough-legged Hawk.** Similar to RL12 but paler individual with pale head and noticeably darker carpal patches. [CO, Feb]

RL14. **Adult male Rough-legged Hawk.** Adult males have grayish backs and upperwing coverts, with some buff and whitish mottling. Tails show numerous bands. Wingtips reach past tail tip. [CO, Dec]

RL15. **Adult male Rough-legged Hawk.** Males usually have multiple dark tail bands and breast equally or more heavily marked than belly. Flanks are always barred. Note feathered legs. Wingtips reach past tail tip. [TX, Nov]

RL16. **Adult male Rough-legged Hawk.** Pale individual with unmarked belly. Dark breast markings form bib. Note legs feathered to toes. Compare to adult Swainson's Hawk in photo SH15. [CO, Mar]

RL17. **Adult female Rough-legged Hawk.** Adult females have belly more heavily marked than breast and single black band on undertail. Leg feathers of adults are heavily marked. Wingtips reach past tail tip. [CO, Dec]

RL18. **Adult male dark-morph Rough-legged Hawk.** Overall black. Legs feathered to toes and black tail has narrow white bands. Some adult males are overall dark brown. [CO, Mar]

RL19. **Adult female dark-morph Rough-legged Hawk.** Overall dark brown. Note white forehead. Legs are feathered to toes. [TX, Dec]

RL20. **Juvenile Rough-legged Hawk.** Juveniles have pale eyes, light heads, lightly streaked breast, and dusky tip to undertail. Juvenile leg feathers are unmarked. Wingtips reach tail tip. [CO, Nov]

RL21. **Juvenile dark-morph Rough-legged Hawk.** Like adult female but with pale eyes and dusky tip to pale undertail. This bird has pale head. Note white forehead and compare to dark-morph juvenile Ferruginous Hawk in photos FH16–17. Wingtips almost reach tail tip. [CO, Jan]

Eagles

Eagles are large dark raptors that have proportionally longer wings than do the smaller buteos, which they resemble in flight. Two species are widespread in North America: Golden Eagle and Bald Eagle. Two other species, the White-tailed Eagle and Steller's Sea Eagle, occur as vagrants.

Golden and Bald Eagles are similar in size; in both species, females are larger than males and northern birds larger than southern ones. Juveniles of all eagles have longer tails and wider wings in their first (juvenile) plumage than they do in later (subadult and adult) plumages.

BALD EAGLE   (*Haliaeetus leucocephalus*)   Photos BE01–19

The Bald Eagle, one of two large brown North American eagles, has five recognizably different plumages that correspond with age: juvenile in its first year, Basic I (White Belly I) in its second, Basic II (White Belly II) in its third, Basic III (Transition) in its fourth, and adult plumage in its fifth plumage when four years old. Sexes are alike in plumage, but females are larger. **Adults** are unmistakable with white heads, tails, and undertail coverts and dark brown bodies and wing coverts. In their first three plumages, **Juvenile to Basic II**, Bald Eagles are mostly dark with a variable amount of whitish mottling, often forming local patches. In flight they all show white axillaries and diagonal white lines on underwings. **Juveniles** have dark brown heads and dark cere and beak. Eye color is dark brown. Back and upperwing coverts are brown and fade after some time to paler brown that contrasts with dark brown uppersides of flight feathers. Breasts are dark brown and usually contrast with bellies that are brown in fresh plumage but fade with time to tawny, or, in rare cases, creamy. Upperside of dark brown tail shows a variable amount of whitish marbling, often as an oval spot, but can be whitish with dark edges and wide terminal bands or entirely dark. Underside of tail is usually whitish with dark edges and wide dark terminal band but can be mostly dark, a juvenile character (see photo BE10). The next two plumages are similar and are characterized by whitish bellies and backs. **Basic I** (White Belly I). Dark head shows wide pale superciliary lines and grayish cere and beak. Iris is usually pale brown, but can be whitish. Back and upperwing coverts are same-color dark brown as flight feathers with whitish spotting, usually heavier on back forming a whitish triangle. Dark brown breast contrasts with whitish belly, which has a variable amount of dark markings. Tail is variable like those of juveniles on upperside but always whitish with dark edges and dark narrow terminal band on underside. Molt of secondaries is not complete, half are long retained juvenile ones and half are shorter replacement ones, combination results in saw-tooth trailing edges of wings (see photos BE05 and 06). **Basic II** (White Belly II). Similar to Basic I but with yellowish beak and cere, pale yellow to whitish eyes, and whitish cheeks, which accentuate narrow dark eye-lines. Molt of secondaries is complete, so trailing edges of wings are smooth (see photo BE04); however, a single long juvenile secondary can be retained on one or both wings. **Basic III**. Transition plumage similar to that of adult. Yellowish beak shows dirty spots, head usually has Osprey-like dark eyeline and some dark spots, and tip of tail usually shows black terminal band, but tail is highly variable in this plumage and can be like those of first three plumages. Body and wing coverts usually show some white spots. Some eagles may still show white triangle on back. First **adult** plumage at four years of age may have hint of dark eye-lines and black tips on some tail feathers.

Best field marks of flying birds are head and neck projection more than half the tail length and wings held level or with slight dihedral. Note also the white axillary patches and a diagonal whitish line on each underwing of flying eagles in first three plumages. Adults with white head and tail are unmistakable.

Similar Species:
Golden Eagles (photos GE01–10) are similar to non-adult Bald Eagles in being mostly dark and may show white on underwings and tail but have legs feathered down to toes, and on flying eagles the head and neck project less than one-half the length of tail and white on underwing is never on coverts or axillaries but is restricted to bases of flight feathers. Golden Eagles with white on base of tail usually have white extending to edges. All but juvenile Goldens show a tawny bar on median coverts of each upper-wing.

Range: Occurs throughout North America but is only common in summer near large bodies of water, mainly in Florida, Chesapeake Bay, coastal Maine through the Maritime Provinces, Great Lakes, Boreal lakes from w. Ontario through coastal British Columbia, most of Alaska coastal areas south along Pacific coast to n. California, the greater Yellowstone area, with scattered local breeders occurring elsewhere. In winter they are more widespread as northern birds move south into continental U.S.

Measurements:
      Length: 70–90 cm (79); 27–35 in. (31)
   Wingspan: 180–225 cm (203); 71–89 in. (80)
     Weight: 2.0–6.2 kg (4.3); 4.4–13.6 lb. (9.5)

BE01. **Adult Bald Eagle.** White head and tail are distinctive. Head and neck project more than half the length of tail. [CO, Jan]

BE02. **First plumage adult Bald Eagle** White head may show Osprey-like eye stripe, and tails may have narrow dark band on tip. This bird has unusually whitish undersides. [MO, Dec]

BE03. **Basic III (Transition) Bald Eagle.** The molt of adult body and wing coverts is not complete; undertail coverts still mostly immature. Note also whitish on secondaries, bill and eye color not lemon yellow, dark Osprey-like eye stripe, and black terminal band on tail. [CO, Jan]

BE04. **Basic II (White Belly II) Bald Eagle.** Whitish belly contrasts with dark breast, similar to Basic I. Aged by smooth trailing edge of wings. Tail same as Basic I. White axillaries and a diagonal white band on each underwing occur in first three plumages. [CO, Jan]

BE05 **Basic I (White Belly I) Bald Eagle.** Whitish belly contrasts with dark breast. Saw-tooth trailing edge of wing due to a mixture of longer juvenile and shorter replacement secondaries. Tail same as juvenile, but shorter. [MO, Dec]

BE06. **Basic I (White Belly I) Bald Eagle.** Similar to BE05 but overall whiter. White belly contrasts with dark brown breast. Saw-tooth trailing edge of wing is noticeable. White axillary patches and a diagonal white band on each underwing occur in first three plumages. [MO, Dec]

BE07. **Basic I Bald Eagle.** White area on back. Coverts same color as flight feathers, but with much whitish mottling. Basic II pattern same. [MO, Dec]
BE08. **Juvenile Bald Eagle.** Two-toned pattern on uppersides: paler back and upperwing coverts contrast with darker flight feathers. [MO, Dec]

BE09. **Juvenile Bald Eagle.** Juveniles have tawny bellies and pointed-tip secondaries that are all the same length. White axillary patches and a diagonal white band on each underwing occur in first three plumages. Whitish tail has dark edges and terminal band. [CO, Feb]

BE10. **Juvenile Bald Eagle.** Juveniles have tawny bellies and pointed-tip secondaries that are all the same length. This juvenile has a darker tail with irregular white markings; some birds have tails that appear all dark. [MO, Dec]

BE11. **Adult Bald Eagle.** White head, undertail coverts, and tail and brown body are distinctive. Note also bright lemon yellow beak. [MO, Dec]

BE12. **First plumage adult Bald Eagle.** Note hint of Osprey-like eye stripe and lack of whitish feathers on body. White tails often have a dark terminal band. [VA, Aug]

BE13. **Basic III (Transition) Bald Eagle.** Resemble adults. Some dark feathers on head and neck and some white feathers on body and coverts. Beak color is not yet bright lemon yellow. Tail highly variable from juvenile-like to white with black tip. [VA, July]

BE14. **Basic II (White Belly II) Bald Eagle.** Like Basic I in photo BE15 but with pale yellow eye, cere, and much of beak. Basic II eagles also have whitish cheeks and dark eye-lines. [AK, Mar]

BE15. **Basic I (White Belly I) Bald Eagle.** Dark breast contrasts with whitish belly. Note wide pale superciliary line, brown eye, dark cheek, and grayish cere and beak. [CO, Jan]

BE16. **Juvenile Bald Eagle.** The tawny belly is noticeable on older juveniles as color fades more than darker breast color. Note dark eye, cere, and beak and wide dark terminal band and dark edges on tail. [CO, Feb]

BE17. **Juvenile Bald Eagle.** Recently fledged juvenile showing dark crown, eye, cere, and beak. Note color of dark upperwing coverts and compare to older juvenile in photo BE19. [VA, July]

BE18. **Basic II (White Belly II) Bald Eagle.** White triangle on back and dark brown heavily spotted upper-wing coverts are shared by Basic I birds. This individual aged by yellow cere and base of beak and whitish malar setting off dark eye stripe. [AK, Feb]

BE19. **Juvenile Bald Eagle.** Note brown back lacking any white and tawny upperwing coverts that contrast with darker secondaries. [MO, Dec]

GOLDEN EAGLE   (*Aquila chrysaetos*)   Photos GE01–10

The Golden Eagle, one of two large brown North American eagles, occurs in hilly and mountainous areas of the West and across northern Canada. It has three recognizably different plumages: adult, subadult (Basic I-III), and juvenile. Females average larger than males. Goldens soar with wings in a slight to medium dihedral and glide with wings in a modified dihedral. Wingtips reach tail tip on perched adult and almost reach on subadults and juveniles. In all plumages, head, body and coverts are dark brown, except for golden crown and nape and rufous undertail coverts. **Adult's** body and coverts often look mottled because of contrast between fresh dark feathers and old faded ones. Tawny median upperwing coverts form a bar across each upperwing. Undersides of flight feathers have grayish mottling, except for wide tip, which results in a dark terminal band on the trailing edge of wings. Adult females have a different tail pattern from males: three narrow wavy grayish bands on males and a wide irregular gray blob across center of tail on females. **Juveniles** appear darker than older eagles and more uniform in color as all body feathers are same age. Median upperwing coverts are dark tawny-brown and do not contrast with upperwing to form bars. Undersides of flight feathers usually have white bases forming patches on underwings (also sometimes form smaller ones on upperwings). Amount of white in patches is variable from none to lots. Otherwise flight feathers are uniform grayish, lacking pale gray marbling and dark tips of other plumages. Tail has white base and dark brown tip, with sharp, well-defined boundary between them. **Subadults** are eagles from one to four years old (Basic I-III). **Basic I** eagles are much like juveniles except that about half of the secondaries have been replaced; new ones are shorter and have dark tips like those of adults. Also new tail has less white at base and more ill-defined boundary between white and dark. Body and wing coverts begin to appear mottled because of mixture of new darker feathers and old faded ones. Tawny bar on upperwing is now discernible. **Basic II and III** eagles are almost identical to adults, including appearing mottled, and have adult-like secondaries but still show some white in base of tail and, in some cases, at base of flight feathers. Adult plumage is acquired at four or five years of age. **Note:** The amount of white in underwings of non-adult Golden Eagles is **not** an age character. Many juveniles do not show any white there and some Basic III eagles show lots.

Best field marks are golden hackles, legs feathered to toes, head projection less than half tail length, and tri-colored bill and cere.

Similar Species:
Bald Eagle (photos BE01–19) non-adults can also appear mostly dark brown but in flight have head that projects more than half tail length, show white axillaries on underwings, and have dark edges to white undertails. Balds usually have cere and beak the same color, have feathered legs, and lack tawny upperwing bars of subadult and adult Goldens.
Dark-morph buteos are much smaller and show silvery flight feathers on underwings when in flight and lack golden napes.
Range: Breeds in hilly and mountainous areas of all of Alaska, most of Canada and the western U.S., and northern Maine. Most northern birds move south for winter.
Measurements:
  Length: 70–84 cm (77); 27-33 in. (30)
  Wingspan: 185–220 cm (200); 72-87 in. (79)
  Weight: 3–6.4 kg (4.5); 6.6–14 lb. (10)

GE01. **Adult male Golden Eagle.** Overall dark brown. Head projects less than half tail length. Note pale grayish markings on flight feathers and dark tips, which form a band on trailing edge. Males have narrow gray tail bands. [WY, Oct]

GE02. **Adult female Golden Eagle.** Overall dark brown. Head projects less than half tail length. Note pale grayish markings on flight feathers and dark tips, which form a band on trailing edge. Females have wide irregular grayish tail bands. [CO, Mar]

GE03. **Subadult male Golden Eagle.** Older subadult, most likely Basic III. Note white in tail. Underwing pattern like adults, but some birds may still show white on base of flight feathers. Males have narrow gray tail bands. [CO, Mar]

GE04. **Adult male Golden Eagle.** Note golden hackles and tawny bar on upperwing. Males have narrow gray tail bands. [WY, Oct]

GE05. **Subadult (Basic II) Golden Eagle.** Two year old plus eagle shows retained longer and pointed juvenile and new shorter and blunter subadult secondaries. Tail now lacks sharp well-defined border between white and dark. [CO, Jan]

GE06. **Juvenile Golden Eagle.** Juveniles have uniformly colored flight feathers lacking dark band on tips and tail with white base and dark tip and well-defined border between them. Amount of white on bases of flight feathers is highly variable. [CO, Jan]

GE07. **Juvenile Golden Eagle.** Individual with little white on undersides of flight feathers. Juveniles have uniformly colored flight feathers lacking dark band on tips and tail with white base and dark tip and well-defined border between them. [CO, Apr]

GE08. **Adult Golden Eagle.** Note golden hackles and tawny bar across wing coverts. Wingtips reach tail tip. Eye is golden or tawny. Cere is yellow, and base of bill is horn-colored and tip is dark; compare to non-adult Bald Eagles in photos BE14–19. [CO, Jan]

GE09. **Subadult Golden Eagle.** Like adult but with white in base of tail. Wingtips fall just short of tail tip. Cere is yellow, and base of bill is horn-colored and tip is dark; compare to non-adult Bald Eagles in photos BE14–19. [CO, Dec]

GE10. **Juvenile Golden Eagle.** Lacking tawny bar on wing coverts. Wingtips fall just short of tail tip. Cere is yellow, and base of bill is horn-colored and tip is dark. Juvenile tail with white base and dark tip. [CO, Jan]

CRESTED CARACARA *(Caracara plancus)*

Photos CA01–07

The Crested Caracara, a large, long-legged, unusual falconid, has two recognizably different plumages: adult and juvenile. Females are slightly larger than males. They soar with wings flat and often flap when soaring but glide and flap on cupped wings. **Adults** have black crown, white face and neck, white upper back narrowly barred black, black lower back, white upper breast narrowly barred black, black lower breast, belly, and flanks, and white tail coverts. White tail has wide subterminal and many narrow black bands. Black wings show whitish primary panels. Large patch of bare face skin and cere are usually bright orange, and legs are bright yellow. **Juveniles** are patterned like adults but are colored brown and buffy instead of black and white and have buffy streaking in place of black barring on upper breast and upper back. Their primary wing panels are whitish, their face skin and cere are pink, and their legs are dull gray. **Basic I birds (Subadults)** are almost like adults in plumage, but dark coloration is dark brown rather than black, necks are buffy rather than white, and barring on upper back and breast is less well defined. Fine rufous streaking is noticeable on crowns of close birds. Face skin varies from pink to dull yellow, and legs are yellow.

Best field marks are large crested head, long neck, and long legs.

Similar Species: Caracaras are quite different from all other raptors; nothing is similar.

Range: Prairies of central Florida, coastal prairies of Texas barely into western Louisiana, and a small area in south central Arizona.

Measurements:

Length: 54–60 cm (58); 21–24 in. (23)
Wingspan: 118–132 cm (125); 46–52 in. (49)
Weight: 800–1300 g (1006); 1.8–2.8 lb (2.2)

CA01. **Adult Caracara**. Black body, black wings with white primary panels, long white neck, and large head with dark crown are distinctive. Note orange face skin and cere, yellow legs, and fine blackish barring on white upper breast and upper back. [TX, Dec]

CA02. **Adult Caracara**. Black body, black wings with white primary panels, long white neck, and large head with dark crown are distinctive. Note orange face skin and cere and fine blackish barring on white upper back. [TX, Dec]

CA03. **Basic I Caracara.** One year old birds are like adults but are dark brown rather than black, with buffy necks and less well defined barring on upper back and breast. Face skin varies from pink to dull yellow, and legs are yellow. [TX, Dec]

CA04. **Juvenile Caracara.** Brownish body, brownish wings with pale primary panels, long buffy neck, and large head with dark crown are distinctive. Note pinkish face skin and cere, pale gray legs, and brown streaking on buffy breast. [TX, Dec]

CA05. **Adult Caracara.** Large head with black crown, white face and neck, orange face skin and cere, and pale beak are distinctive. Body is mostly black, except for white upper breast that is finely barred black. Legs are yellow. Wingtips just reach tail tip. [TX, Dec]

CA06. **Basic I Caracara.** One year old birds are similar to adults but are dark brown rather than black, with buffy necks and less well defined barring on upper back and breast. Face skin varies from pink to dull yellow, and legs are yellow. [TX, Dec]

CA07. **Juvenile Caracara.** Large head has brown crown, buffy face and neck, pinkish face skin and cere, and pale beak are distinctive. Body is mostly brown, but buffy with brown streaking on upper breast and upper back. Wingtips just reach tail tip. [TX, Dec]

Falcons

Falcons are small to medium-sized raptors with long pointed wings. All are in the genus *Falco*. Eight species occur in North America. Five are regular breeders; two, the Northern Hobby and Eurasian Kestrel, occur as vagrants; and the last, the Aplomado Falcon, bred formerly and is now being reintroduced.

Falcons are characterized by bare orbital skin (eye-ring), which is usually the same color as the cere. All have notched beaks used to kill their vertebrate prey by severing the spinal column at the neck. They are active predators, but most will, on occasion, eat carrion.

While falcons are not close taxonomically to other diurnal raptors, they nevertheless share many characteristics, including sharp, curved talons; hooked beaks; excellent eyesight; and both predatory and scavenging habits.

AMERICAN KESTREL   (*Falco sparverius*) Photos AK01–06

The American Kestrel, our smallest falcon, is widespread and common over most of North America. It has two recognizably different plumages: adult male and adult female, but juvenile males differ somewhat from adult males until post-juvenile molt in fall. Females are slightly larger than males. They soar with wings flat and hover regularly. Wingtips of perched kestrels fall short of tail tip. **Adult males** have rufous backs (barred on lower half) and tails, latter with wide black subterminal band; blue-gray upperwing coverts; and pale rufous wash on underparts. Tail pattern of males is highly variable from almost completely rufous to gray, white, and black with no rufous, with many variations in between. **Juvenile males** are similar to adult males but have heavily streaked breasts, completely barred backs, and dark streaks in crown patch (not usually seen). **Adult females** have reddish-brown backs and upperwing coverts that are barred with dark brown, buffy underparts with heavy rufous-brown streaking, and reddish-brown tails with dark brown banding. **Juvenile females** are essentially identical to adults.

Best field marks are white cheeks with two bold black mustache marks, pale underwings of flying kestrels, and row of white dots on trailing edge of males' underwings.

Similar Species:
Merlins (photos M01–14) are similar in proportions but are overall darker, have a much heavier flight, and show dark underwings. They lack the obvious bold mustache marks of kestrels.
Peregrines (photos P01–13) are much larger, with wider, proportionally longer wings, and a single wide mustache mark. Wingtips reach or almost reach tail tips on perched Peregrines.
Range: Common and widespread throughout North America, except for arctic tundra. Northern kestrels move south for winter.
Measurements:
       Length: 22–27 cm (24); 8–10 in. (9)
   Wingspan: 52–61 cm (56); 20–24 in. (22)
      Weight: 97–150 g (116); 3.4–5.3 oz (4.1)

AK01. **Adult male American Kestrel.** Males have pale underwings, spotted rufous underparts, and rufous tail with black subterminal band. Note row of white spots on trailing edges of wings. [CO, Mar]

AK02. **Adult female American Kestrel.** Females have pale underwings, streaked underparts, and reddish-brown tail with dark brown bands. Black mustache marks are visible when falcon is near. Juvenile female is identical. [CO, Apr]

AK03. Adult male American Kestrel. Adult males have rufous backs with barring on lower half, some spotting on belly, blue-gray wing coverts, and rufous tail. Note bold black mustache marks on white cheeks. Wingtips fall short of tail tip. [CO, Mar]

AK04. **Female American Kestrel.** Females have reddish-brown backs and wing coverts with dark brown barring and reddish-brown tail with dark brown bands. Note bold black mustache marks on white cheeks. Wingtips fall short of tail tip. [TX, Oct]

AK05. Juvenile male American Kestrel. Similar to adult male but breast has blackish streaking. Note bold black mustache marks on white cheeks. [CT, Oct]

AK06. **Juvenile male American Kestrel.** Similar to adult male but back is completely barred. Note bold black mustache marks on white cheeks. [CT, Oct]

APLOMADO FALCON (*Falco femoralis*) Photos AF01–04

The Aplomado Falcon, a medium-sized falcon, was extirpated from the Southwest in the recent past but is now being reintroduced in Texas. It has two recognizably different plumages: adult and juvenile. Females are larger than males. They have proportionally longer tails than do other falcons. Wingtips of perched falcons extend three-fourths of way to tail tip. All show bold head pattern of gray crown, pale buffy superciliaries that join to form a V on hind-neck, narrow dark eye-lines, pale buffy cheeks, long narrow black mustache marks, and pale buffy throat. **Adults** have grayish back and upperwing coverts, buffy underparts crossed by a black belly band, and rufous lower belly, leg feathers, and undertail coverts. Long black tail has six or more narrow white bands. Belly band shows fine white barring. Females have fine black streaks on breast; males usually lack them. Adults have yellow ceres and eye-rings. **Juveniles** are similar to adults, but have dark brown back and upperwing coverts, wide dark streaks on breast, and pale-buffy lower belly, leg feathers, and undertail coverts. Dark brown belly band shows narrow pale buffy streaking. Long dark brown tail has six or more narrow pale buffy bands. Juveniles have bluish ceres and eye-rings.

Best field marks are bold head pattern, dark belly band, and pale band on trailing edges of dark wings. Long tail is also distinctive.

Similar Species:
American Kestrels (photos AK01–06) are smaller with shorter wings and tails, show two black mustache marks, have pale underwings, and lack dark belly band.
Merlins (photos M01–14) are smaller, have completely streaked underparts, shorter wings and tails, and lack bold head pattern, dark belly band, and pale band on trailing edges of wing.
Peregrines (photos P01–13) are larger, with wider wings and single wide mustache mark, and lack dark belly band.
Prairie Falcons (photos PF01–07) are larger and paler and lack dark belly band.
Range: Formerly bred in s. Texas, w. Texas, and s. New Mexico and Arizona. Now being reintroduced in s. Texas.
Measurements:
>   **Length:** 35–45 cm (40); 14–18 in. (16)
>   **Wingspan:** 78–102 cm (89); 31–40 in. (35)
>   **Weight:** 208–460 g (328); 8.4–16 oz (11.6)

AF01. **Adult female Aplomado Falcon.** Bold head pattern, dark belly band, narrow pale band on trailing edge of dark underwings, and long dark tail with many narrow whitish bands. Females show fine dark streaking on breast. [Mexico, Feb]

AF02. **Juvenile Aplomado Falcon.** Similar to adults with bold head pattern but upperparts are dark brown. Note narrow pale band on trailing edge of wings. [Mexico, Oct]

AF03. **Adult male Aplomado Falcon.** Bold head pattern, dark belly band, rufous leg feathers, and long dark tail with narrow white bands are distinctive. Males usually lack dark streaking on breast. Adults show fine white barring on dark belly band. [Mexico, Feb]

AF04. **Juvenile (female) Aplomado Falcon.** Similar to adult but breast heavily streaked with dark brown, upperparts dark brown, and lower belly, leg feathers, and undertail coverts are pale buffy. Juvenile females have heavier breast streaking than do juvenile males. Wingtips fall somewhat short of tail tip. [Mexico, Oct]

MERLIN   (*Falco columbarius*)   Photos M01–14

The Merlin, a small dark falcon, is relatively rare wherever it occurs. It has six recognizably different plumages: adult male and adult female of three races: Taiga, Prairie, and Black. Females are separably larger than males. Wingtips of perched Merlins fall short of tail tip. They soar with wings flat and do not hover. **Adult males** have blue-gray backs and upperwing coverts and whitish heavily streaked underparts, with a rufous wash on sides of breast and leg feathers. Their black tails have gray banding. **Adult females** have dark brown backs and upperwing coverts and buffy heavily streaked underparts. Their dark brown tails have buffy banding. **Juveniles** differ from adult females only in Taiga Merlins; they lack adult females' grayish cast to uppertail coverts. Compared to **Taiga Merlins, Prairie Merlins** are overall paler and **Black Merlins** are overall darker. Taiga Merlins are darkest in East and palest in West.

Best field marks are dark underwings (not so dark on Prairie Merlins) and lack of a bold mustache mark.

Similar Species:
American Kestrels (photos AK01–06) are smaller, appear paler, show rufous on back and tail, and have two bold black mustache marks.
Peregrines (photos P01–13) are larger, have relatively longer wings, and have a wide mustache mark. Their wingtips reach or almost reach tail tip when perched.
Sharp-shinned Hawk (photos SS01–07) can appear similar to perched female and juvenile Taiga Merlins but have rounded heads, yellow or red eyes, and equal-width dark and pale tail bands.
Prairie Falcons (photos PF01–07) can appear similar to Prairie Merlins but are larger, have a distinct narrow mustache mark, and show dark axillaries on underwings.
Range: Taiga Merlins breed in the boreal forest from Alaska to Newfoundland and winter from southern U.S. southward. Prairie Merlins breed on the northern Great Plains and winter there and southward. Black Merlins are resident in the wet temperate forests of coastal British Columbia and Washington, with some birds moving southward in winter. Some Merlins around the west end of Lake Superior appear much like typical Black Merlins.

Measurements:
Length: 24–30 cm (26); 9–12 in. (10)
Wingspan: 53–68 cm (61); 21–27 in. (24)
Weight: 129–236 g (178); 4.5–8.3 oz (6.5)

M01. **Adult male Taiga Merlin.** Underparts are heavily streaked with noticeable rufous washes on upper breast, leg feathers, and undertail coverts. Note heavily marked underwing and black tail with pale bands. Mustache mark noticeable but not bold. [NJ, Sept]

M02. **Adult male Taiga Merlin.** Upperparts are blue-gray, and black tail has narrow blue-gray bands. [NJ, Sept]

M03. **Adult male Prairie Merlin.** Streaking on underparts is narrow. Note rufous wash on leg feathers and black tail with wide whitish bands. Mustache mark faint or absent. [Sask, July]

M04. **Adult female Taiga Merlin.** Heavily streaked underparts lack rufous wash. Dark tail has three narrow pale bands. Mustache mark noticeable but not bold. Juveniles appear identical. [MN, July]

M05. **Adult female Taiga Merlin.** Upperparts are dark brown and dark brown tail has narrow whitish-buffy bands. Juveniles are almost identical. [MN, July]
M06. **Adult female Prairie Merlin.** Back color is paler brown and pale tail bands are wider. Juveniles same. [Sask, July]

M07. **Adult female Prairie Merlin.** Like Taiga adult female and juvenile, but streaking is paler brown and narrower. Mustache mark faint or absent. Juveniles are not separable by plumage. [Sask, July]

M08. **Adult male Taiga Merlin.** Back is blue-gray. Note rufous leg feathers and faint mustache mark. Wingtips do not reach tail tip. This bird is especially dark. [WI, July]

M09. **Adult female or juvenile Taiga Merlin.** Note dark brown back, faint mustache mark, and heavily streaked underparts. Adult females and juveniles are almost identical. Wingtips do not reach tail tip. [TX, Jan]

M10. **Adult male Prairie Merlin.** Back is pale blue-gray. Note faint mustache mark. Wingtips do not reach tail tip. [CO, Feb]

M11. **Female Prairie Merlin.** Compared to Taiga Merlins, overall paler, with narrow streaking, fainter mustache mark, and wider pale bands in tail. Juveniles and adult females are identical in plumage. This falcon judged a female by size. [CO, Jan]

M12. **Juvenile male Taiga Merlin.** Darker falcon than M09. Sexed as male by size. [CT, Oct]

M13. **Juvenile male Prairie Merlin.** Back is medium brown, paler than those of Taiga Merlins. Wingtips do not reach tail tip. Note faint mustache mark and wide pale tail bands. Sexed as male by size. [CO, Dec]

M14. **Female Black Merlin.** Extremely dark Merlin, with dark cheek, no pale superciliary line, heavily streaked underparts, faint narrow, incomplete tail bands, and barring on undertail coverts. Sexed as female by size; adult females and juveniles are identical. Adult males (not shown) have dark bluish cast to upperparts. [TX, Dec]

PRAIRIE FALCON   (*Falco mexicanus*)   Photos PF01–07

The Prairie Falcon, a large pale falcon, is resident in arid hills and mountains of the West. It has two subtly different plumages: adult and juvenile. Females are larger than males. Prairies soar with wings flat. All are brownish above and whitish below and have square, blocky pale heads with a white area behind each eye and long narrow dark mustache marks. Median underwing coverts of females are more heavily marked than those of males. Wingtips of perched falcons fall short of tip of long tail. **Adults** show pale barring on back feathers and upperwing coverts and dark spotting on whitish underparts. Adults have yellow ceres, eye-rings, and legs. **Juveniles'** backs lack pale barring and appear darker than those of adults. Underparts of juveniles are streaked. Cere and eye-ring are bluish; legs are pale yellow.

Best field marks are dark axillaries and wing coverts on underwings of flying falcons and square head with white marks directly behind eyes on perched falcons.

Similar Species
Peregrines (photos P01–13) are similar in size but are overall darker with wider dark mustache mark and uniformly dark underwings (but note juvenile Tundra Peregrine in photos P05, 11–12). Wingtips reach or almost reach tail tip on perched Peregrines.
Prairie Merlins (photos M10–11, 13) are similar in coloration but are much smaller, have faint mustache marks, and lack dark axillaries.
Swainson's Hawk juveniles (photo SH21) can appear similar when perched but their wingtips almost reach tail tips.
Range: Breed in drier hills and mountains of West barely into southern Canada. In winter some falcons move south, west, and east. Casual in eastern U.S.
Measurements:
> **Length:** 37–47 cm (41); 14–18 in. (16)
> **Wingspan:** 90–113 cm (102); 36–44 in. (40)
> **Weight:** 420–1100 g (686); 0.9–2.1 lb. (1.6)

PF01. **Adult male Prairie Falcon.** Overall pale, with black axillaries. Adults have spotted underparts. Males have rather pale median underwing coverts. Note narrow dark mustache mark. [WY, June]

PF02. **Adult female Prairie Falcon.** Overall pale, with black axillaries and median coverts. Adults have spotted underparts. Females have darker median underwing coverts. Note narrow dark mustache mark. [WY, June]

PF03. **Juvenile Prairie Falcon.** Overall pale, with black axillaries. Juveniles have streaked underparts. Note narrow dark mustache mark. [WY, July]

PF04. **Adult Prairie Falcon.** Adults have brown backs with pale buffy bands across many feathers. Note narrow dark mustache mark. [WY, May]

PF05. **Juvenile Prairie Falcon.** Juveniles have uniformly dark brown backs and upperwing coverts. White area behind eye separates juveniles from juvenile Tundra Peregrines (compare to photo P05). Note narrow dark mustache mark. [CO, July]

PF06. **Adult Prairie Falcon.** Adults have brown backs with pale buffy bands across many feathers and yellow ceres. Note square, blocky head with narrow dark mustache marks and white area behind eyes. Wingtips fall short of tail tips. [WY, June]

PF07. **Juvenile Prairie Falcon.** Juveniles have bluish ceres, streaked underparts, and dark brown backs lacking pale barring; back appears darker than those of adults. Note square, blocky head with narrow dark mustache marks and white area behind eyes. Wingtips fall short of tail tips. [CO, Feb]

GYRFALCON *(Falco rusticolus)*   Photos GY01–08

The Gyrfalcon, our largest falcon, breeds in the arctic tundra of Alaska and northern Canada. It has six recognizably different plumages: adult and juvenile of three color morphs, white, gray, and dark. Females are larger than males. Gyrfalcons soar with wings flat. Wingtips of perched falcons reach two-thirds down to tail tip. **Adults** have yellow ceres, eye-rings, and legs. **Gray- and dark-morph adults** have dark backs and upperwing and uppertail coverts with pale barring and pale underparts with streaking on breasts, spotting on belly, and barring on flanks. Intergrades exist in a cline between gray and dark morphs. **White-morph adults** have white backs and upperwing coverts, usually with black barring and often have some black spotting on white underparts. Underwings are all white except for black wingtips. **Juveniles** have bluish ceres, eye-rings, and legs and completely streaked underparts. **Gray- and dark-morph juveniles** have dark backs and upperwing and uppertail coverts. Upperparts of gray-morph juveniles appear scalloped owing to pale feather edging. **White-morph juveniles** have brown backs and upperwing coverts with wide white feather edges and a variable amount of streaking on white underparts. White-morph falcons lack mustache marks; all others have faint narrow mustaches, except for darkest falcons with all dark cheeks.

Best field marks are two-toned underwings on gray and dark morph falcons and all white body and underwings, latter with black wingtips.

Similar Species:
Peregrines (photos P01–13) are smaller, have uniformly dark underwings, narrower wings with more pointed wingtips, and thicker mustache marks. Their wingtips reach or almost reach tail tip on perched falcons.
Prairie Falcons (photos PF01–07) are smaller and paler and have dark axillaries on underwings and white areas behind eyes.
Goshawk adults (photos G01–03) can appear similar when flapping but have two-toned upperwings, dark heads with wide white superciliaries, and barred underwings.
Range: Breed in Arctic tundra across North America. Some falcons, usually juveniles, move south for winter barely into northern U.S.
Measurements:
      Length: 50–61 cm (52); 19–24 in. (22)
    Wingspan: 110–130 cm (121); 43–51 in. (47)
      Weight: 1000–2100 g (1420); 2.2–4.6 lb. (3.1)

GY01. **Adult gray-morph Gyrfalcon**. Typical adult with streaked breast, spotted belly, and barred flanks. Note pale underwings and long, wide-based tail. [AK, June]

GY02. **Adult white-morph Gyrfalcon**. Adults have white upperparts and upper tail with a variable amount of black barring. Note lack of dark mustache mark. [NewF, Feb]

GY03. **Juvenile dark-morph Gyrfalcon**. Gray- and dark-morph juveniles have two-toned underwings: dark coverts contrast with paler flight feathers. Note wide tail and lack of wide mustache marks. [CT, Mar]

GY04. **Adult white-morph Gyrfalcon**. Adults have white upperparts and upper tail with a variable amount of black barring. This falcon is in its second year, notice the retained juvenile upperwing coverts and blue cere. Adults have yellow ceres. [NewF, Mar]

GY05. **Adult gray-morph Gyrfalcon**. Dark gray back and upperwing coverts with white to pale gray cross barring. White underparts are marked with barring on flanks, and spots on breast and belly. Cere is yellow on adults. [WA, Feb]

GY06. **Juvenile white-morph Gyrfalcon**. Juveniles have dark brown back and upperwing coverts with white feather edges. Tail is white with dark bands like those of adults. Cere is blue on juveniles. Wingtips fall rather short of tail tip. [NewF, Apr]

GY07. **Juvenile gray-morph Gyrfalcon.** Juveniles have whitish underparts heavily streaked and bluish ceres. Their grayish-brown backs lack cross barring (not shown). Note faint mustache mark of gray morph. Wingtips fall short of tail tip. [MN, Dec]

GY08. **Juvenile dark-morph Gyrfalcon.** Upperparts are uniformly dark grayish-brown and underparts are mostly dark with some whitish streaking. Note narrow dark mustache mark. Wingtips fall quite short of tail tip. This individual is rather pale headed. [CT, Mar]

PEREGRINE *(Falco peregrinus)*

The Peregrine, a large, long-winged falcon, is widespread throughout North America but is nowhere common. It has two recognizably different plumages: adult and juvenile. There is also some plumage variation among the three North American races, Anatum, Tundra, and Peale's, as well as much intergrading among them. Females are separably larger than males. Peregrines soar on flat wings. **Adults** have dark heads, appearing "hooded", with dark mustache marks and white cheek patches, blue-gray upperparts, and whitish underparts with clear breasts and barred bellies and flanks, as well as yellow ceres, eye-rings, and legs. **Anatum adults** have darker heads with wider mustache marks and smaller cheek patches and often show a strong salmon wash on underparts. **Tundra adults** are overall paler, with narrower mustache marks and large white cheek patches and less salmon on underparts. **Peale's adults** lack salmon wash and are more heavily marked on underparts, especially showing dark streaking on breasts and dark streaks in white cheek patches. They often exhibit a grayish bloom to upperparts. **Juveniles** have dark brown upperparts and buffy underparts with dark streaking, as well as bluish ceres and eye-rings; their leg color varies from greenish to yellow. **Anatum juveniles** have dark heads with wide mustache marks and small buffy cheek patches and chevron-shaped dark markings on leg feathers. **Tundra juveniles** are quite different, with much paler heads, some almost blond, much narrower mustache marks, narrower dark streaking on buffy underparts, and dark streaks on leg feathers. **Peale's juveniles** are darker; their underparts vary from heavily streaked to almost uniformly dark, their white cheeks show dark streaks, and their leg feathers are dark with pale edges.

Best field marks are prominent dark mustache marks, uniformly dark underwings on flying falcons, and wingtips reaching or almost reaching tail tip on perched ones.

Similar Species:
Prairie Falcons (photos PF01–07) are similar in size but are overall paler, show black axillaries on underwings, and their wingtips fall short of tail tip when perched. Note white marks behind eyes on perched Prairies.
Gyrfalcons (photos GY01–08) are larger, have wider, more rounded tip on wings, often show two-toned underwings in flight, and wingtips fall quite short of tail tip when perched.
Range: Tundra Peregrines breed on the arctic tundra from western Alaska to Greenland and migrate south to Latin America for winter. Peale's Peregrines breed in the Aleutian Islands and along the Pacific Northwest coast from British Columbia south to Oregon and are mainly sedentary, but with some movement southward in winter. Anatum Peregrines breed in subarctic Alaska and Canada and throughout the West. A variety of Peregrines now breed in the East, owing to reintroduction efforts.
Measurements:
>**Length:** 37–46 cm (42); 14–18 in. (16)
>**Wingspan:** 94–116 cm (104); 37–46 in. (41)
>**Weight:** 453–952 g (704); 1–2.1 lb. (1.6)

P01. **Adult female Anatum Peregrine.** Adults have dark heads that appear hooded, clear breasts, and barred bellies. Note wide mustache mark of anatum adult; they usually have more salmon wash on breast. Peregrines have uniformly dark underwings. Adult females are usually more heavily barred on underparts than are males.
[Sask, July]

P02. **Adult Tundra Peregrine.** Adults have dark heads that appear hooded, clear breasts, and barred bellies. Note narrow mustache of Tundra adults. Peregrines have uniformly dark underwings. [NJ, Oct]

P03. **Adult Anatum Peregrine.** Upperparts of adults are blue-gray. Note dark head with wide dark mustache mark that appears hooded. [Sask, July]

P04. **Juvenile Anatum Peregrine.** Juveniles are heavily streaked on underparts. Note wide mustache mark and dark head of Anatum juveniles. Peregrines have uniformly dark underwings. [CA, July]

P05. **Juvenile Tundra Peregrine.** Juveniles are heavily streaked on underparts. Note pale head and narrow mustache mark of Tundra juvenile. [NJ, Oct]

P06. **Adult Anatum Peregrine.** Adults have dark heads, blue-gray upperparts, clear breasts, and barred bellies. Note wide mustache mark of Anatum adult; they usually have more salmon wash on breast. Wingtips reach tail tip. [Sask, July]

P07. **Adult Anatum Peregrine.** Variant with completely dark cheek and rather heavily marked breast. [AZ, Dec]

P08. **Adult Tundra Peregrine.** Adults have dark heads, blue-gray upperparts, clear breasts, and barred bellies. Note narrow dark mustache mark and white cheek of Tundra adults. [NC, Dec]

P09. **Adult Peale's Peregrine.** Similar to Anatum adult with wide mustache mark and barred belly but has dark streaking on breast and cheeks and lacks salmon wash on underparts. [CA, Aug]

P10. **Juvenile Anatum Peregrine.** Juveniles have dark heads, dark brown backs, and heavily streaked underparts. Wingtips fall a bit short of tail tip on juveniles. Note wide mustache mark and chevron markings on leg feathers. [NM, Aug]

P11. **Juvenile Tundra Peregrine.** Tundra juveniles have pale heads, narrow mustache marks, and vertical streaks on leg feathers. Typical pale-headed individual. [LA, Jan]

P12. **Juvenile Tundra Peregrine.** Tundra juveniles have pale head and narrower mustache marks. Typical darker-headed individual. Wingtips reach tail tip. [MD, Oct]

P13. **Juvenile Peale's Peregrine.** Peale's juveniles are darker, with dark underparts that may show some pale streaking. Note mottled cheek and wide mustache mark. Leg feathers are almost completely dark, with narrow buffy edges. [NY, Nov]

Vagrants

CRANE HAWK   (*Geranospiza caerulescens*)   Photo CH01

The Crane Hawk, a medium-sized, long-legged, dark gray Neotropical raptor, has occurred once in s. Texas as a vagrant from Mexico. It has two plumages: adult and juvenile. Females are larger than males. Wingtips of perched birds barely reach base of long tail. They glide on cupped wings. **Adults** are overall dark grayish, with red eye, long orangish legs, and long black tail with narrow white bands. A curved row of white spots is visible near wingtips on flying birds. **Juveniles** are similar to adults but are more brownish-black, show white superciliaries and cheeks, have white barring on underparts, and wider white tail bands. Eye color is pale, not red.

Best field marks are long orangish legs, reddish eyes (of adults), and curved row of white spots on wingtips on flying birds.

Similar Species: No other raptors in s. Texas are similar.
Range: Widespread but local resident throughout Latin America. Resident in Mexico in lowland coastal areas north to s. Tamaulipas. An adult was observed and photographed at Santa Ana NWR in the Rio Grande Valley of s. Texas.
Measurements: (No data available.)

CH01. **Adult Crane Hawk.** Overall dark grayish. Note red eye, long orangish legs, and long black tail with narrow white bands. [Colombia, Aug]

WHITE-TAILED EAGLE   (*Haliaeetus albicilla*)

Photos WE01–03

The White-tailed Eagle, a large Eurasian sea eagle similar to the Bald Eagle, is a vagrant to Alaska and eastern North America. It has five age-related plumages, corresponding to similar age Bald Eagles. Sexes are alike in plumage, but females are larger. Wingtips of perched eagles reach tail tip. They soar with wings flat. **Adults** are similar to adult Bald Eagles with all brown body and wing coverts, but their heads are creamy, not white, with color extending onto and lacking a sharp line of contrast with brown upper breast and upper back. Undertail coverts are brown, and short white tail has wedge-shaped tip. Non-adults (**Juveniles and Basic I and II**) are similar to same age Bald Eagles but have more uniform underparts and show less white on axillaries and underwing coverts. Non-adults show black triangular marks on tips of white tails, resulting in a "spiked" appearance. **Basic III** eagles are usually dark overall except for mostly whitish tail and yellowish beak.

Best field marks to separate from Bald Eagles are wider wings, shorter, more wedge-shaped tail, dark undertail coverts of adults, and more uniform underparts of non-adults.

Similar Species:
Bald Eagle (photos BE01–19) juveniles are similar, but have longer tail and narrower wings with only six "fingers" on wingtips (White-tailed show seven) and have breast noticeably darker than belly. Black band on tail tips are square, not triangular.
Steller's Sea Eagles (photos SE01–03) are larger and have much larger beak, more pointed wingtips, and longer, even more wedge-shaped tails.
Range: Vagrant in Alaska, especially the Aleutian Islands and Kodiak Island, and Eastern North America. Most vagrants are juveniles, but a pair of adults have bred on Attu Island in the eastern Aleutian Islands. A probable sight record of a White-tailed Eagle was recorded in New York.
Measurements:
       Length: 77–92 cm (84); 30–36 in. (33)
   Wingspan: 208–247 cm (231); 82–97 in. (91)
      Weight: 3.1–6.9 kg (4.8); 6.8–15.2 lb. (10.6)

WE01. **Adult White-tailed Eagle.** Head is creamy, lacking clear-cut border with dark body. Undertail coverts are dark, and short white tail is wedge-shaped. Note seven "fingers" on wingtips. [Japan, Feb]

WE02. **Juvenile White-tailed Eagle.** Darkish breast same color as belly. Whitish axillaries and white lines on underwings are similar to those on non-adult Bald Eagles. Note triangular black mark on tip of each tail feather and seven "fingers" on wingtips. [Japan, Feb]

WE03. **Adult White-tailed Eagle.** Head is creamy, lacking clear-cut border with dark body. Wingtips reach tail tip. [Japan, Feb]

STELLER'S SEA EAGLE   (*Haliaeetus pelagicus*)

Photos SE01–03

The Steller's Sea Eagle, the world's largest eagle, is a vagrant to Alaska from northeastern Asia. It has five age-related plumages, corresponding to similar aged Bald Eagles, but the first three plumages are almost identical. Females are larger than males. Wingtips of perched eagles almost reach tail tip. They soar with wings in a strong dihedral. All have huge yellow to orange beaks, long, wedge-shaped tails, and blackish bodies and flight feathers. **Adults** have white upperwing and underwing coverts, leg feathers, and tail coverts and show grayish streaking on crown and nape. **Juveniles** and other **immatures (Basic I and II)** have dark coverts and leg feathers and whitish mottling in dark axillaries and underwing coverts and a white patch at base of primaries on each underwing. They usually show a variable amount of black mottling or black tips or both on white tails.

Best field marks are large size, huge yellow or orange beaks, long deeply wedge-shaped tails, and pointed wingtips.

Similar Species:
Bald Eagles (photos BE01–19) are smaller and have much smaller beaks and more rounded wingtips and lack wedge-shaped tails.
White-tailed Eagles (photos WE01–03) are smaller and have much smaller beaks, more rounded wingtips and shorter, less wedge-shaped tails
Range: Vagrant to Alaska, with records from Kodiak Island, Attu and Unalaska in the Aleutian Islands, and the Pribilofs. Also recorded on Midway Island in Hawaiian chain. An adult has been associating with an adult Bald Eagle for several years near Juneau. Normal range is east coast of Siberia; occurs in winter south to the Japanese island of Hokkaido.
Measurements:
      Length: 85–105 cm (95); 33–41 in. (37)
    Wingspan: 220–245 cm (232); 87–96 in. (91)
      Weight: 5–9 kg (7); 11–20 lb. (15)

SE01. **Adult Steller's Sea Eagle.** Distinctive with huge orangish beak, blackish body and flight feathers contrasting with white underwing coverts and leg feathers, and long white wedge-shaped tail. [Japan, Feb]

SE02. **Juvenile Steller's Sea Eagle.** Non-adults are mostly black overall, with huge yellowish beak and long white wedge-shaped tail. Usually show whitish axillaries and patches on undersides of primaries and black in white tail. [Japan, Feb]

SE03. **Adult Steller's Sea Eagle.** Note huge orangish beak, pied plumage, and long wedge-shaped tail. [Japan, Feb]

ROADSIDE HAWK   (*Buteo magnirostris*)   Photos RH01–03

The Roadside Hawk, a small, accipiter-like Neotropical buteo, occurs in s. Texas as a vagrant from Mexico. It has two recognizably different plumages: adult and juvenile. Females are somewhat larger than males. Northern Mexican birds usually lack rufous in primaries. Wingtips of perched birds fall far short of tail tip. **Adults** have gray-brown heads and backs and solid brown breast forming bib (but sometimes with a few whitish streaks) and barred belly. Eyes are pale lemon yellow. **Juveniles** are similar to adults but show heavy pale streaking in bib, short creamy superciliaries, and orangish-yellow eyes and have more and narrower tail bands.

Best field marks are equal width dark and light tail bands, paddle shaped wings, and buffy U of uppertail coverts.

Similar Species:
Gray Hawk juveniles (photos GH02, 04, 06) have bold face pattern, dark eyes, and whiter U above tail base.
Broad-winged Hawk juveniles (photos BW04–06) have more pointed wingtips, show a dark band on trailing edge of relatively unmarked underwings, and lack buffy U above tail. Adult Broad-wings have dark eyes.
Hook-billed Kite adult females (photos HB02, 05, 06) are similar with paddle-shaped wings but have white eyes, large hooked beak, completely barred underparts, and rufous in primaries (but a few Roadsides may show rufous in primaries).
Range: Widespread resident throughout Latin America from Mexico south. Vagrant to the U.S. Recorded a few times along the Mexican border in lower Rio Grande valley of Texas.
Measurements:
        Length: 33–38 cm (36); 13-15 in. (14)
   Wingspan: 72–79 cm (75); 28–31 in. (30)
      Weight: 230–440 g (318); 8–15 oz (11)

RH01. **Adult Roadside Hawk.** Note paddle-shaped wings, lemon yellow eye, dark bib, and equal-width dark and light tail bands. Primaries may or may not have as much rufous as this bird shows. [Mexico, Oct]

RH02. **Adult Roadside Hawk.** Note gray-brown head and back, lemon yellow eye, and equal-width dark and light tail bands. Wingtips fall far short of tail tip. [Mexico, Feb]

RH03. **Juvenile Roadside Hawk.** Like adult but with short pale superciliary, pale streaking on breast, more and narrower tail bands, and orangish eye. [Mexico, Oct]

RED-BACKED BUZZARD   *(Buteo polysoma)*

Photo RB01

The Red-backed Buzzard, a South American buteo, has occurred in south-central Colorado mated with a Swainson's Hawk. Wingtips of perched birds extend beyond tail tip. They soar with wings in a strong dihedral. **Adults** are similar to adult White-tailed Hawks but with completely gray upperwing coverts, lacking chestnut shoulder patches. Some adults have rufous backs (females?); others have gray ones. Underwings appear uniform white; coverts and flight feathers are same color.

Similar Species:

White-tailed Hawk adults (photos WT10–11) are similar but have rufous shoulder patches and wingtips that extend far beyond tail tip noticeable on perched adults and show two-toned underwing in flight; gray flight feathers contrast with whitish coverts.
Range: South America. Adult female from an unknown source has been mated with a male Swainson's Hawk for many years in south-central Colorado. Where it migrates to in winter is unknown.

Measurements:

      Length: 41–44 cm; 16–17 in.
    Wingspan: 115–119 cm; 45–46 in.
      Weight: 500–700 g; 1.1–1.5 lb.

RB01. **Adult female Red-backed Buzzard.** Note gray head, completely gray upperwing coverts, and rufous back. Wingtips extend past tail tip. Compare to adult White-tailed Hawks in photos WT10–11. [CO, Sept]

HAWAIIAN HAWK   (*Buteo solitarius*)   Photos HH01–02

The Hawaiian Hawk, a small buteo, is the only resident diurnal raptor on the Island of Hawaii. It has four recognizably different plumages, adults and juveniles of light and dark color morphs. Females are larger than males. They soar with wings flat, and regularly hunt on the wing, but do not hover. Wingtips of perched birds fall somewhat short of tail tip. Light- and dark-morph birds occur with equal frequency. **Dark-morph adults** are uniformly dark brown on head, body, coverts, and tail, except for silvery undersides of flight feathers and whitish undertail coverts. **Dark-morph juveniles** are similar to adults but with pale rufous streaking on underparts and underwing coverts. **Light-morph adults** have brown head and upperparts and whitish underparts with dark streaking. **Light-morph juveniles** have entirely pale creamy heads and creamy underparts with little or no streaking.

Similar Species: No other raptors on the Island of Hawaii are similar.
Range: Resident only on the big island, Hawaii. Has occurred as a vagrant on other islands.
Measurements: (No data available.)

HH01. **Adult light-morph Hawaiian Hawk.** Adults have dark heads, backs, and upperwing coverts and white underparts with dark streaking. [HA, June]

HH02. **Adult dark-morph Hawaiian Hawk.** Overall dark brown, except for whitish undertail coverts. Wingtips fall short of tail tip. [HA, June]

COLLARED FOREST-FALCON
(*Micrastur semitorquatus*) Photo FF01

The Collared Forest-Falcon, a medium-sized, long-legged and short-winged Neotropical forest raptor, has occurred once in s. Texas as a vagrant from Mexico. It has two light-morph plumages: adult and juvenile. Females are larger than males. Wingtips of perched birds barely reach base of long tail. **Adults** have dark brown crowns, napes, backs, and upperwing coverts and white throats, sides of face, and underparts. White on sides of face extends onto lower part of nape to form a half collar. Long dark tail has three narrow white tail bands. Eyes are dark. **Juveniles** are similar to adults but are paler brown above and are buffy and barred below.

Best field marks are long legs, short wings, and white half collar.

Similar Species: No other raptors in s. Texas are similar.
Range: Widespread but local resident throughout Latin America. Resident in Mexico in heavily forested areas north to s. Tamaulipas. A light-morph adult was observed at Bentsen State Park in the Rio Grande Valley of s. Texas.
Measurements: (No data available.)

FF01. **Adult Collared Forest-Falcon.** Upperparts dark brown and underparts white. Note white semi-collar. Note short wings that barely reach base of tail. Long black tail has narrow white bands. [Guatemala, ?]

COMMON KESTREL   (*Falco tinnunculus*)   Photos CK01–05

The Common Kestrel, a medium-sized falcon, occurs as a vagrant to North America from Eurasia. It has four recognizably different plumages: adult and juvenile of male and female; however, juveniles of both sexes are similar to adult female. Females are slightly larger than males. Like American Kestrels, they regularly hover. Powered flight is with soft, fluttery, shallow wingbeats. Wingtips of perched kestrels fall short of tail tip. All have pale head with a single noticeable narrow dark mustache mark below each eye and show two-toned pattern on upperwings: dark primaries and primary coverts contrast with paler brown secondaries and secondary coverts. **Adult males** have gray heads and tails and chestnut backs and upperwing coverts. **Adult females** have reddish-brown backs and upperwing coverts that are marked with dark brown wide triangles. Reddish-brown to gray tails have many narrow dark brown bands and one wide subterminal dark brown band. Uppertail coverts are usually grayish with black shaft streaks, often with black barring. **Juvenile** plumage of both sexes is subtly different from that of adult female.

Best field mark is two-toned pattern of upperwings and fluttery flight.

Similar Species:
American Kestrels (photos AK01–06) are smaller and more compact, show two bold mustache marks, and have uniform colored upperwings.
Range: Widespread across Eurasia; vagrant to North America, with a handful of records from both coasts.
Measurements:
       Length: 29–38 cm (34); 11–15 in. (13)
    Wingspan: 68–82 cm (76); 27–32 in. (30)
       Weight: 127–280 g (180); 4.5–9.9 oz (6.7)

CK01. **Adult male Common Kestrel.** Adult males have gray heads and gray tails with one black band. Note pale underwings, single narrow dark mustache stripe, and long wedge-shaped tail. [Israel, Sept]

CK02. **Adult male Common Kestrel.** Adult males have gray heads and tails and chestnut backs. Note two-toned upperwing: chestnut coverts contrast with dark brown primaries. [India, Nov]

CK03. **Adult female Common Kestrel.** Adult females have brown heads and backs and two-toned upperwings: medium brown coverts contrast with dark brown primaries. Juveniles are similar. [Israel, Sept]

CK04. **Adult male Common Kestrel.** Adult males have gray heads and tails and chestnut backs and upperwing coverts. Note single narrow dark mustache mark. Wingtips fall short of tail tip. [Israel, Mar]

CK05. **Adult female Common Kestrel.** Adult females have brown backs with dark brown barring. Note single narrow dark mustache mark. Wingtips fall short of tail tip. Juveniles are similar. [Israel, Mar]

NORTHERN HOBBY   (*Falco subbuteo*)   Photos H01–03

The Northern Hobby, a medium-sized falcon, is a vagrant to North America from Eurasia. It has two recognizably different plumages: adult and juvenile. Females are larger than males. Often appears swift-like when gliding, owing to depressed wings. Wingtips of perched falcons reach or extend just beyond tail tip. Dark heads appear hooded with two narrow black mustache marks. **Adults** have dark blue-gray backs and upperwing coverts, heavily streaked whitish underparts, and rufous leg feathers and undertail coverts. **Juveniles** are similar but have brownish backs and upperwing coverts, buffy leg feathers and undertail coverts, and a wide pale band on tail tip.

Best field marks are sickle-shaped wings and dark body and underwings on flying Hobbys, and dark hooded head with two dark mustache marks and wingtips reaching tail tip on perched ones.

Similar Species:
Peregrines (photos P01–13) are larger, with wider mustache stripes, wider wings, and a heavier flight.
Merlins (photos M01–14) are smaller, have shorter wings, and lack well-defined mustache marks.
Range: Widespread in Eurasia; vagrant to Alaska, where there exist a handful of records.
Measurements:
       Length: 29–32 cm (30); 11–13 in. (12)
    Wingspan: 74–83 cm (78); 29–33 in. (31)
       Weight: 141–325 g (237); 5–11.5 oz (8.9)

H01. **Adult Northern Hobby.** Long narrow wings, rufous leg feathers and undertail coverts, and narrow mustache marks are distinctive. Streaks extend from throat to vent. [Israel, Sept]

H02. **Juvenile Northern Hobby.** Like adult but leg feathers and undertail coverts are buffy. Note wide pale terminal tail band. Streaks extend from throat to vent. [Israel, Sept]

H03. **Adult Northern Hobby.** Note dark hood with two finger-like mustache marks, heavy streaking, and rufous leg feathers and undertail coverts. Wingtips extend just beyond tail tip. [Israel, Apr]

Raptor Identification Problems

1. Sharp-shinned vs. Cooper's Hawks

The Sharp-shinned Hawk and Cooper's Hawk are structurally different and do not overlap in size. In the hand, plumage and proportional differences easily distinguish them. Yet, they are notoriously difficult to distinguish under field conditions; this might well be the toughest raptor field ID problem in North America. However, it is possible to accurately identify them in many cases using a combination of the field marks presented below. Accurate field identification usually requires a good look at the hawk.

The following table summarizes their differences.

Character	Sharp-shinned Hawk	Cooper's Hawk
Head and neck projection	Shorter	Longer
Length of tail	Shorter	Longer
Tip of tail	Square or somewhat rounded	Very rounded
Terminal white tail band	Narrow	Wide
Wrist in glide	Pushed far forward	Pushed forward less
Wing in soar	Flat	Flat or dihedral
Belly of juvenile	Usually heavily marked	Lightly marked
Back of juvenile	Dark, few pale marks	Dark, many pale marks
Legs	Stick-like	Stout

SS01. **Adult Sharp-shinned Hawk.** Note short neck and head and short, square-tipped tail with thin white terminal band. [MN, Sept]

C02. **Adult Cooper's Hawk.** Glides with wrists somewhat forward; head projects far beyond wrists. Outer tail feathers are shorter than others. Long tail has rounded tip. [CO, Apr]

SS02. **Juvenile Sharp-shinned Hawk.** Note short neck and head, streaking on underparts extends onto belly, barring on flanks, and short square-tipped tail. [MN, Sept]

C03. **Juvenile Cooper's Hawk.** Head and tail are longer than those of Sharpies. Belly streaking is sparse or absent. Note long tail with shorter outer tail feathers and wide white terminal band. [NJ, Oct]

SS03. **Juvenile Sharp-shinned Hawk.** Head barely projects beyond wrists in glide. Note square-tipped tail with straight narrow white terminal band. [NJ, Oct]

C04. **Juvenile Cooper's Hawk.** Head projects far beyond wrists in glide. Note rounded corners and wide pale band on tip of tail. [NJ, Oct]

SS05. Adult Sharp-shinned Hawk. Crown is same color as back, without line of contrast. Head always rounded. Tail is square-tipped and shows only a narrow white terminal band; outer tail feathers are same length as others. [CA, Jan]

C05. Adult Cooper's Hawk. Crown is darker than back, with sharp line of contrast with paler nape. When hackles are raised, rear of head appears square and eye closer to beak. Outer tail feathers are shorter than others on long tail. Cooper's often sit on poles in open areas, particularly in the West. [NJ, Oct]

SS06. Juvenile Sharp-shinned Hawk. Small rounded head has brighter yellow eye and narrow superciliary line. Streaking of underparts extends onto belly. Legs are stick-like. Tail lacks wide white band; outer tail feathers are same length as others. [NJ, Oct]

C06. Juvenile Cooper's Hawk. Head shows square nape (hackles raised) and duller yellow eye closer to beak than to nape. Fine dark brown streaking is sparse on belly. Legs are stout, not stick-like. Tail feathers have wide white bands; outer tail feathers are noticeably shorter; tip appears rounded. [NJ, Oct]

2. Pale square primary panels on upperwings of flying buteos

Three of the widespread species of North American buteos show a square pale panel on the upperside of the primaries. However, only one species, the Red-tailed Hawk, has pale primary coverts, and then only in juvenile plumage. Pale primaries and coverts on upper wings is a diagnostic field mark for juvenile Red-tails.

Ferruginous Hawks show a pale primary panel on upperwings in all plumages; Rough-legged Hawks only in juvenile plumage. Both species have dark primary coverts. Swainson's Hawk and Broad-winged Hawk do not show pale primary panels on upperwings. However, sometimes juvenile Broad-winged Hawks show a pale trapezoidal panel on underwings. Pale primary panels of all Red-shouldered Hawks are crescent-shaped and appear quite different. Pale trapezoidal primary panels on underwings are shared by juvenile Red-tailed, Broad-winged, and Rough-legged Hawks and all Ferruginous Hawks and are thus not a diagnostic field mark (see problem 3).

The table below summarizes which species by age show large pale panels on uppersides of primaries and coverts.

Species	Age	Primaries	Primary coverts
Red-tailed Hawk	Ad	Dark	Dark
Red-tailed Hawk	Juv	**Pale**	**Pale**
Rough-legged Hawk	Ad	Dark	Dark
Rough-legged Hawk	Juv	**Pale**	Dark
Ferruginous Hawk	All	**Pale**	Dark
Red-shouldered Hawk	All	**Pale crescent**	Dark
All others	All	Dark	Dark

RT19. **Juvenile Eastern Red-tailed Hawk.** Best field mark for all juvenile Red-tails is pale outer half of upperwing, including both primaries and primary coverts. Note brown tail with many narrow dark brown tail bands, subterminal not wider. [NY, Apr]

RT26. **Juvenile Harlan's Red-tailed Hawk.** Shows typical juvenile Red-tail pattern of pale outer wing, including primary coverts, that contrast with darker inner wing. [CO, Nov]

FH04. **Adult Ferruginous Hawk.** Whitish primary panel does not include dark primary coverts. Note chestnut upperwing coverts. Uppertail is rufous and white. Note narrow black tips on outer primaries. [CO, Jan]

FH05. **Juvenile Ferruginous Hawk.** Whitish primary patch does not include dark primary coverts. Back and upperwing coverts are dark brown. White uppertail coverts have large black spots and form a white U at base of uppertail. [CO, Jan]

RL10. **Juvenile Rough-legged Hawk.** Juveniles show whitish patch on uppersides of primaries; primary coverts are dark. Uppertail is white on base with dusky band on tip. Uppertail coverts are also white. [CO, Dec]

RL05. **Adult female Rough-legged Hawk.** Adults lack pale primary patches on upperwings. Note dark subterminal band on uppertail and brownish back. [CO, Jan]

3. Pale primary panels on back-lighted underwings of flying buteos

Five of the widespread species of North American buteos show a pale panel on the undersides of the primaries when the wing is back-lighted. However, in one species, the Red-shouldered Hawk, this panel is crescent-shaped. On the other four species, Red-tailed, Broad-winged, Ferruginous, and Rough-legged Hawks, this panel is trapezoid-shaped.

Ferruginous Hawks show this pale primary panel on underwings in all plumages; Red-tailed, Broad-winged, and Rough-legged Hawks show it only in juvenile plumage. Swainson's Hawks and adult Red-tailed, Broad-winged, and Rough-legged Hawks do not show pale primary panels on underwings.

Pale trapezoidal primary panels on underwings are not a diagnostic field mark, as this character is shared by juvenile Red-tailed, Broad-winged, and Rough-legged Hawks and all Ferruginous Hawks.

Species	Age	Panel shape
Red-tailed Hawk	Ad	None
Red-tailed Hawk	Juv	Trapezoid
Rough-legged Hawk	Ad	None
Rough-legged Hawk	Juv	Trapezoid
Ferruginous Hawk	All	Trapezoid
Red-shouldered Hawk	All	**Crescent**
Broad-winged Hawk	Ad	None
Broad-winged Hawk	Juv	Trapezoid
Swainson's Hawk	All	None

RS06. **Juvenile Red-shouldered Hawk.** Typical juvenile with crescent-shaped pale wing panels visible on back-lighted wings. Note streaked underparts, dark tail with narrow pale bands, and wings pushed forward in soar. [TX, Oct]

RS02. **Adult Red-shouldered Hawk.** Typical adult with whitish crescent-shaped wing panel visible on back-lighted wings. Note rufous underwing coverts and white tail bands. [TX, Jan]

RT17. **Juvenile Red-tailed Hawk.** Back-lighted wings show pale trapezoidal primary panels. Note dark patagial mark and pale brown tail with many narrow dark bands. Belly band not diagnostic. [MN, Sept]

BW06. **Juvenile Broad-winged Hawk.** Back-lighted wings show pale trapezoidal primary panels. Note plain underwings with dark border. Tail shows narrow dark bands with subterminal band wider. [MN, Sept]

FH06. **Juvenile Ferruginous Hawk.** Back-lighted wings show pale trapezoidal primary panels. Overall whitish with few markings. Note dark flank patches, long tapered wings, lack of dark patagial mark, and whitish unbanded base of tail. [CO, Jan]

RL10. **Juvenile Rough-legged Hawk.** Back-lighted wings show pale trapezoid primary panels. Note dark carpal batches, dark band on undertail, and solid belly band. [CO, Jan]

4. White scapular markings on backs of perched buteos

Whitish patches on scapulars have been considered as a diagnostic field mark for Red-tailed Hawks. But some Swainson's Hawks, particularly non-adults, can also show whitish scapulars and resemble juvenile Red-tails in tail pattern and other characters.

All Red-tailed Hawks, regardless of age or race, show a variable amount of whitish marking on the scapulars, which appear as a pale V on the backs of perched birds. All but dark-morph Western Red-tails and the darkest Harlan's have this character.

Since some Swainson's Hawks share this character, it is not a diagnostic field mark for Red-tailed Hawks, except in areas and at times when Swainson's would not be present. If Swainson's Hawks are possible, then different field marks will serve to distinguish these species. If the Red-tail is an adult, then it will have a rufous or whitish (Harlan's) tail. However, if it is a juvenile, then whether or not the wingtips reach the tail tip is the diagnostic field mark; it reaches on Swainson's and falls quite short of Red-tails.

RT27. **Adult Eastern Red-tailed Hawk.** Back brownish with whitish mottling on scapulars. Adult Red-tails show rufous tail. [TX, Jan]

RT32. **Adult Krider's Red-tailed Hawk.** Back brownish with whitish mottling on scapulars; markings heaviest on this color morph. Identified by pale rufous tail. [LA, Dec]

RT38. **Juvenile Red-tailed Hawk.** Back brownish with whitish mottling on scapulars. Wingtips do not reach tail tip. Brown tail has multiple narrow dark brown bands, all the same width. [TX, Dec]

SH21. **Juvenile (one year old) light-morph Swainson's Hawk.** Back brownish with whitish mottling on scapulars. Wingtips reach tail tip. Dark subterminal band is the same width or barely wider than other dark tail bands. [WY, June]

RT42. **Juvenile Harlan's Red-tailed Hawk.** Back brownish with whitish mottling on scapulars. Wingtips fall somewhat short of tail tip. Light brown tail has wide dark brown chevron-shaped bands. Probable light-morph. [MO, Nov]

SH24. **Juvenile light-morph Swainson's Hawk.** Back brownish with whitish mottling on scapulars. Wingtips almost reach tail tip. Juvenile's dark subterminal band same width or barely wider than other dark tail bands. Juveniles recently fledged show pale feather edges. [CO, Sept]

5. Golden Eagle vs. non-adult Bald Eagle

Adult Bald Eagles are easy to identify because of their white heads and tails. But in their first three plumages, they appear much like Golden Eagles. However, with a good look, one can easily separate non-adult eagles.

Perched Golden Eagles show golden crown and nape and three colors on cere and beak. Whereas Bald Eagles have dark crowns and one color on cere and beak. White-belly Balds show wide whitish to buffy superciliary lines.

In flight non-adult Golden Eagles may show white at bases of flight feathers on underwings but have dark axillaries and wing linings. Conversely, non-adult Balds have white axillaries (wing pits) and usually show white lines across their wing linings; these are not visible when entire linings are white.

Tails of non-adult Goldens are white at base with dark tips; tails of non-adult Balds can appear similar, but always have dark edges on their tails.

In flight, the heads of Golden Eagles project less than half the tail length, whereas those of Balds project more than half.

BE19. **Juvenile Bald Eagle.** Note all dark head and beak and cere same color. Tawny upperwing coverts contrast with darker secondaries. Basic I and II Balds show whitish mottling on back. [MO, Dec]

GE10. **Juvenile Golden Eagle.** Note golden hackles. Cere is yellow, and base of bill is horn-colored and tip is dark. White at base of tail extends to edge. [CO, Jan]

BE09. **Juvenile Bald Eagle.** Non-adult Bald Eagles have white axillaries (wing pits) and a diagonal white band on each underwing. Whitish tail has dark edges. Head projects more than half tail length. [CO, Feb]

GE06. **Juvenile Golden Eagle.** Non-adult Golden Eagles have dark axillaries and underwing coverts. Whitish patches are restricted to base of flight feathers. Whitish tail has white edges. Head projects less than half tail length. [CO, Jan]

BE05 **Basic I (White Belly I).** Non-adult Bald Eagles have white axillaries and diagonal white bands on underwings. Tail has dark edges. Head projects more than half tail length. [MO, Dec]

GE05. **Basic II (Subadult) Golden Eagle.** Non-adult Golden Eagles have dark axillaries and wing linings. Whitish patches are restricted to base of flight feathers. Head projects less than half the tail length. [CO, Jan]

6. Perched juveniles with streaked underparts, pale superciliaries, and dark malar stripes

Many raptors in juvenile plumage appear similar when perched. They show brownish upperparts and pale underparts that are heavily streaked. They all show pale superciliary lines and dark malar stripes as well.

Shown here are juveniles of the No. Goshawk and five buteonines, Broad-winged, Red-shouldered, Gray, Common Black, and Swainson's Hawks.

The tail pattern is the main field mark to distinguish the species, but other marks are useful.

A common problem is distinguishing juvenile Broad-wings and Red-shouldereds. Broad-wings have short legs, unmarked secondaries, and pale brown uppertails with dark brown bands, whereas Red-shouldereds have long legs, pale banding on secondaries, and dark brown tails with narrow pale banding.

Wing tips fall quite short of tail tip on Broad-winged, Red-shouldered, and Gray Hawks and on No. Goshawks; but almost reach on Common Black and do reach on Swainson's Hawks.

Juvenile Goshawks show irregular tail banding with white highlights; Common Black Hawks have whitish tails with wavy narrow dark bands.

Gray Hawks have a distinctive dark eye-line.

BW11. **Juvenile Broad-winged Hawk.** Shows pale superciliary, dark malar stripe, and streaked underparts. Note short legs, unmarked secondaries, and light brown uppertail with narrow dark brown bands. [NJ, Oct]

RS13. **Juvenile Red-shouldered Hawk.** Shows pale superciliary, dark malar stripe, and streaked underparts. Note long legs, pale bands on folded secondaries, dark tail with narrow pale bands, and rufous wash at base of tail. [TX, Jan]

GH06. **Juvenile Gray Hawk.** Shows pale superciliary, dark malar stripe, and streaked underparts. Note striking face pattern, especially dark eye-line, and barred leg feathers. Outer dark bands on long brown tail are progressively wider. [Mexico, Jan]

G07. **Juvenile Goshawk.** Shows pale superciliary, dark malar stripe, and streaked underparts. Note tawny bars on upperwing coverts, dark spots on undertail coverts, and irregular bands and white highlights on uppertail. [NJ, Nov]

BH06. **Juvenile Common Black Hawk.** Shows pale superciliary, dark malar stripe, and streaked underparts. Note that wide dark malar stripe extends onto neck and white tail with wavy dark bands. [AZ, July]

SH23. **Juvenile Swainson's Hawk.** Shows pale superciliary, dark malar stripe, and streaked underparts. Note dark patches on sides of upper breast. Wingtips almost reach tail tip. [CO, Sept]

7. Dark-morph Rough-legged Hawk vs. Harlan's Hawk

Adult dark-morph Rough-legged Hawks, particularly adult males, are very similar in appearance to adult Harlan's Hawks, especially those with banded tails. Both are blackish with black and white tails and have heavily marked pale undersides of flight feathers.

The subtle differences in wing shapes and position will serve the experienced observer to distinguish flying birds. Harlan's usual adult tail lacks banding. The dark band on the tail tip is wider with a clear cut border on Rough-legs and narrower and less well defined on Harlan's.

On perched birds, wingtips reach the tail tips of perched Rough-legs, but do not reach on perched Harlan's. Rough-legged's legs are feathered to the toes, whereas those of Harlan's are bare.

Harlan's usually (but not always) show whitish streaking in breast, and the commonest tail patterns are whitish to grayish and mottled, not banded.

Adult females and juveniles (and a few adult males) are dark brown and show black carpal patches on underwings.

RL17. **Adult male dark-morph Rough-legged Hawk.** Overall blackish. Legs feathered to toes and black tail has narrow white bands. Wingtips reach tail tip. [CO, Mar]

RT36. **Adult Harlan's Red-tailed Hawk.** Overall blackish. Some have underparts with no whitish streaking. Alternate tail pattern: whitish with narrow dark bands, subterminal wider. Wingtips fall somewhat short of tail tip. [CO, Nov]

RL06. **Adult male dark-morph Rough-legged Hawk.** Adult males are blackish. Black tail has wide dark band on tip. Undersides of silvery flight feathers are heavily barred; whitish area on outer primaries is large. [CO, Dec]

RT12. **Adult Harlan's Red-tailed Hawk.** Harlan's are blackish. Undersides of silvery flight feathers are heavily barred; whitish area on outer primaries is small. Variation with banded tail; band on tail tip is narrow. This individual lacks white breast patch. [CO, Nov]

RL07. **Adult female dark-morph Rough-legged Hawk.** Overall dark brownish. Undertail is whitish with wide, sharply defined dark subterminal band. Underwings show black carpal patch. [CO, Dec]

RT11. **Adult Harlan's Red-tailed Hawk.** Typical adult with gray mottled tail and whitish breast patch. Dark band on tail tip is not well defined. [IA, Oct]

8. Perched Juveniles of Red-tailed Hawk, Ferruginous Hawk, and Swainson's Hawk

In western North America, juvenile buteos can appear quite similar and difficult to identify when perched. But close inspection of the back pattern, tail pattern, and wingtip position relative to the tail tips are useful field marks that will serve to identify them.

Juvenile Red-tailed Hawks have mottled backs and pale brown tails with equal-width dark bands. On perched birds, wingtips fall way short of tail tip.

Juvenile Ferruginous Hawks have unmarked dark brown upperparts and darker brown tails with only faint darker banding. Their wingtips fall a bit short of tail tips. Their legs are feathered.

Juvenile Swainson's Hawks have mottled backs and pale brown tails with (usually) equal-width dark tail bands. Their wingtips reach or almost reach tail tips.

RT38. **Juvenile Eastern Red-tailed Hawk**. Red-tails show whitish patches on scapulars. Juvenile's dark subterminal band same width or barely wider than other dark tail bands. Wingtips fall short of tail tip. [TX, Dec]

SH24. **Juvenile light-morph Swainson's Hawk.** Juveniles show pale feather edges, including whitish patches on scapulars. Juvenile's dark subterminal band same width or barely wider than other dark tail bands. Wingtips almost reach tail tip. [CO, Sept]

FH14. **Juvenile Ferruginous Hawk.** Back is dark brown, lacking any whitish mottling or pale feather edges. Note dark eye-line. Tail bands not distinct. Wingtips fall somewhat short of tail tip. [CO, Dec]

RT41. **Juvenile Krider's Red-tailed Hawk.** Like normal juvenile Red-tails but with whitish head and more whitish mottling on back and wing coverts. Wingtips fall short of tail tip. [LA, Feb]

SH21. **Juvenile (one year old) light-morph Swainson's Hawk.** Juveniles just returning from winter grounds still show whitish scapular patches and appear quite pale headed. Wingtips reach tail tip. [WY, June]

FH13. **Juvenile (one year old) Ferruginous Hawk.** Back is dark brown and unmarked. Year-old juvenile shows worn plumage. Legs are feathered almost to toes. [WY, June]

9. Perched dark-morph buteos

Many dark buteos have uniformly dark brown bodies and wing coverts and appear very similar. The best field marks for their separation are their tail patterns, for example, the rufous tail of adult Red-tailed Hawks, the dirty white mottled tail of Adult Harlan's Red-tails, and the gray unmarked tail of adult Ferruginous Hawks. For others with more subtle differences in tail pattern, there are other field marks that also serve to distinguish them, as enumerated below.

Juvenile Red-tails' wingtips fall somewhat short of the tail tip, and their legs are unfeathered.

The wingtips of juvenile and adult female Rough-legs reach the tail tip, and their legs are completely feathered. Note also that they usually have a small whitish area on the forehead.

Juvenile Ferruginous Hawks show a large gape and have wingtips that almost reach the tail tip. Their legs are completely feathered.

Adult Swainson's Hawks have unfeathered legs and white undertail coverts. Wingtips reach the tail tips on perched birds. Non-adult dark-morph Swainson's are streaked and not uniform dark brown.

SH19. **Adult dark-morph Swainson's Hawk.** Overall dark brown, except for whitish undertail coverts. Note dark forehead. Legs are unfeathered. Wingtips reach tail tip. [CO, Sept]

FH16. **Juvenile dark-morph Ferruginous Hawk.** Overall dark brown. Note dark forehead. Legs are feathered to toes. Note large gape. Wing tips fall somewhat short of tail tip. [CA, Nov]

RL20. **Juvenile dark-morph Rough-legged Hawk.** Overall dark brown, except for white forehead. This bird has pale head. Undertail is whitish with dusky tip. Legs are feathered to toes. Note narrow gape. Wingtips almost reach tail tip. [CO, Jan]

RT43. **Juvenile dark-morph Red-tailed Hawk.** Overall dark brown. Some birds have rufous streaking on underparts. Legs are unfeathered. Wingtips fall short of tail tip (not shown). [CO, Nov]

RL18. **Adult female dark-morph Rough-legged Hawk.** Overall dark brown. Note white forehead. Legs are feathered to toes. [TX, Dec]

10. Perched pale-headed buteos

A perched pale-headed buteo in western North America could be one of four species. These species, as well as their age and sex, are listed below, along with the field marks necessary to distinguish them.

Krider's and juvenile Harlan's Red-tailed Hawks are usually pale headed. They have wingtips that fall somewhat short of the tail tip, and their legs are unfeathered. Krider's juveniles usually show a few dark spots suggesting a dark belly band; light Harlan's show distinct belly bands; and adult Krider's lack belly bands.

Juvenile and adult female Rough-legs are usually pale headed and show a dark eyeline. Their wingtips reach the tail tip, and their legs are completely feathered. They both have a wide dark belly band.

One year old juvenile Ferruginous Hawks often have a whitish head. They show a large gape and have completely feathered legs. Their wingtips almost reach their tail tip.

One year old Swainson's Hawks just returned from the winter grounds usually have whitish heads. They have unfeathered legs and streaked underparts. Their wingtips reach their tail tips.

SH22. **Juvenile (one year old) Swainson's Hawk.** Juveniles returning from winter grounds appear quite pale headed before molt. Dark subterminal band same width or barely wider than others. Wingtips almost reach tail tip. [ID, May]

RT41. **Juvenile Krider's Red-tailed Hawk.** Like normal juvenile Red-tails but with whitish head and mottling on back and wing coverts. Whitish underparts show faint belly band. Legs are unfeathered. Wingtips fall short of tail tip. [LA, Feb]

FH13. **Juvenile Ferruginous Hawk.** Year-old juvenile shows worn plumage. Note dark eye-line. Whitish underparts are mostly unmarked. Whitish leg feathers extend almost to toes. [WY, June]

RT42. **Juvenile Harlan's Red-tailed Hawk.** Probable light morph has whitish head. Note whitish markings on upperparts and chevron-shaped tail bands. Wingtips fall short of tail tip. [MO, Nov]

RT32. **Adult Krider's Red-tailed Hawk.** Head is usually whitish. Note whitish spotting on upperparts and pale rufous tail. Wingtips do not reach tail tip. [LA, Dec]

RL17. **Adult female Rough-legged Hawk.** Usually have pale head. Note dark eye-line and whitish underparts with wide dark brown belly band. Legs are feathered to toes. Wingtips reach past tail tip. Juveniles are similar. [CO, Dec]

11. Perched juvenile White-tailed Hawk vs. other buteos

Juvenile White-tailed Hawks are quite different from adult White-tails, being overall dark blackish-brown, with a variable amount of white streaking on the breast. They also show several pale spots on the sides of their dark heads. Their tails are grayish with numerous faint narrow dark banding. When perched, their wingtips extend a ways beyond the tail tip.

In the range of the White-tailed Hawk along the Gulf Coast of Texas, some Swainson's Hawks and Harlan's Red-tailed Hawks could be confused with the similar juvenile White-tailed Hawks. The field marks to distinguish them are enumerated below.

Dark-morph juvenile and Basic I Swainson's Hawks lack face spots and have distinctive tail banding. Their streaking occurs throughout the underparts, and their wingtips barely reach the tail tips when they are perched.

Dark-morph Harlan's, both adults and juveniles, lack pale face spots, and wingtips fall short of tail tip on perched birds. Adult's dirty white tail has distinct mottling or banding, and tail bands in juveniles' tails are distinct.

WT13. **Juvenile White-tailed Hawk.** Dark underparts usually have white breast patch. Note long legs. Dark head shows whitish patches on side of face. Tail has faint narrow dark bands. Wingtips exceed tail tip. [TX, Jan]

SH20.   **Basic I (two year old) dark-morph Swainson's Hawk.** Basic I bird just before molting. Dark underparts have whitish streaking all over. Undertail shows distinct bands. Wingtips reach tail l tip. [ID, May]

WT14. **Juvenile White-tailed Hawk.** Paler juvenile with larger white breast patch, white markings on belly, and whitish leg feathers. Dark head shows whitish spots on side of face. Wingtips exceed tail tip. Tail has faint narrow dark bands. [TX, Jan]

RT37. **Adult Harlan's Red-tailed Hawk.** Overall blackish with whitish markings on face and whitish streaking on breast. Grayish tail shows either heavy mottling or bold banding. Wingtips fall short of tail tip. [CO, Nov]

WT15. **Juvenile White-tailed Hawk.** Dark head shows whitish spots on side of face. Dark upperparts show some faint rufous markings. Wingtips exceed tail tip. Tail has faint narrow dark bands. [TX, Jan]

RT46. **Juvenile Harlan's Red-tailed Hawk.** Dark brown back shows whitish on scapulars. Chevron-shaped dark tail bands are distinct. Wingtips fall somewhat short of tail tip. [CO, Nov]

12. Perched raptors with dark bibs

It has long been part of the birding lore that the dark breast or bib of adult Swainson's Hawks is diagnostic. However, individuals of at least three other species can show this character, namely some adult male light-morph Rough-legged Hawks, non-adult Bald Eagles, and some adult Broad-winged Hawks.

Adult Swainson's Hawks have either rufous or dark brown breasts forming a bib and almost always have a white throat. Their paler bellies vary from unmarked to heavily barred and tails are pale brown with dark brown bands.

Bald Eagles in their first three plumages have dark brown breasts and paler bellies, always lacking the white throat patch. The bellies of juveniles vary from a paler brown to pale tawny; those of Basic I and II are white with a variable amount of dark brown markings. Their undertails are white with dark tip and edges.

Some adult Broad-winged Hawks have uniformly rufous breasts forming a bib, with some barring on flanks and belly. They show a single white band in their dark tails but never show white throat.

Some adult male light-morph Rough-legged Hawks have dark breasts forming a bib and some barring on their flanks, but never show a white throat. Their white tails have one or more dark bands near the tip and their legs are feathered to their toes.

SH14. **Adult light-morph Swainson's Hawk.** Typical adult with rufous breast forming bib and paler unmarked belly. Note whitish throat patch. Pale undertail shows dark banding, subterminal band wider. [CO, Sept]

BE16. **Juvenile Bald Eagle.** Dark breast forms a bib; tawny belly color fades more than darker breast color. Note wide dark terminal band and dark edges on tail and lack of white throat. [CO, Feb]

RL15. **Adult male Rough-legged Hawk.** Dark brown breast markings form bib. Note barred flanks and legs feathered to toes. Pale undertail shows wide dark subterminal band. Wingtips barely exceed tail tip. [CO, Feb]

SH16. **Adult light-morph Swainson's Hawk.** Typical adult with dark brown breast forming bib and heavily barred belly. Note whitish throat patch. Pale undertail shows dark banding, subterminal band wider. [CO, Sept]

BW08. **Adult Broad-winged Hawk.** Variation with solidly marked breast forming bib. Note single wide white band in dark undertail. [TX, Oct]

BE15. **Basic I (White Belly I) Bald Eagle.** Dark breast forms a bib and contrasts with whitish belly. Note wide pale superciliary line, brown eye, dark cheek, and dark beak. Undertail is whitish with dark tip and edges. [CO, Jan]

13. Perched juvenile large falcons

Perched juveniles of the three large falcons are similar in having brownish backs and streaked underparts. Head patterns, mustache marks, and wingtip/tail tip positions are important field marks.

The first problem is separating juvenile Prairie Falcon from juvenile Tundra Peregrine. Both are brown backed with narrowly streaked underparts and narrow dark mustache marks. Prairie Falcons have large block-shaped heads and show a small white area behind each eye. When perched, their wingtips fall quite short of their tail tips. Peregrines show dark behind the eye and juveniles, when perched, have their wingtips reach near to the tail tips.

Separating juvenile Gyrfalcons from Peregrines is the next problem. Gyrs wingtips fall somewhat short of tail tip; whereas those of Peregrines almost reach. Gyrs have relatively small heads and massive chests and narrow dark mustache marks. Peregrines have wide mustache marks.

P12. **Juvenile Tundra Peregrine.** Tundra juveniles have pale head and narrower mustache marks. Typical paler-headed individual. Note dark behind eyes. Wingtips almost reach tail tip. [MD, Oct]

PF07. **Juvenile Prairie Falcon.** Prairies have square, blocky heads with narrow dark mustache marks and a small white area behind each eye. Wingtips fall short of tail tips. [CO, Feb]

P10. **Juvenile Anatum Peregrine.** Anatum juveniles have dark heads with wide dark mustache marks. Wingtips fall a bit short of tail tip on juveniles. Note chevron markings on leg feathers of Anatum juveniles. [NM, Aug]

GY07. **Juvenile gray-morph Gyrfalcon.** Gray-morph Gyrfalcons show faint mustache marks. Wingtips fall quite short of tail tip. [MN, Dec]

P13. **Juvenile Peale's Peregrine.** Peale's juveniles are darkest Peregrines, with dark underparts that may show some pale streaking. Note dark hood, mottled cheek, and wide mustache mark. Wingtips fall just short of tail tip. [NY, Nov]

GY08. **Juvenile dark-morph Gyrfalcon.** Dark-morph juvenile Gyrs have dark underparts with some whitish streaking. Note narrow dark mustache mark and paler crown. Wingtips fall quite short of tail tip. [CT, Mar]

14. Dark-morph Gyrfalcon vs. dark Peregrine in flight

Juvenile Peale's Peregrines and juvenile dark-morph Gyrfalcons appear similar in flight. Both are large, dark falcons with darkly streaked underparts.

But their underwings differ. Those of Peregrines are uniformly dark, whereas those of Gyrfalcons are two-toned: dark coverts contrast with paler flight feathers.

Peregrines have wider mustache marks compared to those of dark-morph Gyrfalcons. Gyrs also have wider base of tails.

P04. **Juvenile Anatum Peregrine.** Peregrines have uniformly dark underwings. Juveniles are heavily streaked on underparts. Note wide mustache mark and dark head of anatum juveniles. [CA, July]

GY03. **Juvenile dark-morph Gyrfalcon.** Mostly dark below with faint whitish streaking. Note two-toned underwings: dark coverts contrast with paler flight feathers. Lacks wide dark mustache mark of dark Peregrines; compare to photo P04. [CT, Mar]

INDEX